T0205623

Hyperspectral Image Fusion

Subhasis Chaudhuri · Ketan Kotwal

Hyperspectral Image Fusion

 Springer

Subhasis Chaudhuri
Ketan Kotwal
Department of Electrical Engineering
Indian Institute of Technology Bombay
Powai, Mumbai
India

ISBN 978-1-4899-9375-5   ISBN 978-1-4614-7470-8 (eBook)
DOI 10.1007/978-1-4614-7470-8
Springer New York Heidelberg Dordrecht London

Printed on acid-free paper

Springer is part of Springer Science+Business Media (www.springer.com)

To
Kulkuli and Shiblu

—SC

My parents

—Ketan

Preface

Hyperspectral imaging is a fairly recent technique that has an excellent potential for applications in many diverse areas including the primary area of remote sensing. A fine spectral sampling allows us to capture the spectral signature with great details. A huge amount of data to represent a single scene, however, imposes difficulties in visualizing the image. This motivated us to develop various techniques for visualizing the hyperspectral data cube. Pixel-level image fusion appeared to be a natural choice for visualization of the entire data cube.

While working on the visualization problem, it was felt that the available literature is quite scattered and very sparse compared to the much matured area of basic image fusion. Hence we embarked on writing a monograph that will serve as a reference book in this area to all students and practitioners of hyperspectral imaging technology. The monograph serves two purposes—it provides an up-to-date reference to all research work, and it also provides implementation details of various visualization techniques with the hope that one should be able to implement them easily. We also present a comparative analysis of various competing techniques to enable researchers to decide which method would be more suitable to them.

As always, the decision to write a monograph comes much later during the tenure of research in the relevant area. During the research period, quite a few papers have been published in various journals and conferences, following which these ideas have been extended further in the monograph. The publications that overlap with the contents of the book are given in the references [87–92, 145, 146]. Needless to say, the copy rights of various figures reproduced from these earlier publications belong to the original publisher. It may also be mentioned that some of the visualization outputs are in color. For various cost implications, we refrain from presenting them in color in the hardcopy version of the monograph. However the readers may opt for the e-book version for illustrations in color.

The book is addressed to a broad readership. We expect graduate students and researchers in this area to find this book very useful. Application engineers in the remote sensing area will also find this monograph helpful. We have attempted to make the book as self-contained as possible. However, familiarity of readers with basics of image processing and statistical parameter estimation will help in better appreciating the contents of this book.

We look forward to any comments or suggestions from the readers.

Mumbai, February 2013   Subhasis Chaudhuri
                                                                                                  Ketan Kotwal

Acknowledgments

The authors are highly indebted to Dr. R. Shanmuganathan at the Indian Institute of Technology Jodhpur, for his help in developing the contents of Chap. 6. Various comments and suggestions received from Prof. V. Rajbabu and Prof. S. N. Merchant of IIT Bombay are gratefully acknowledged. Thanks are also due to Prof. A. N. Rajagopalan of IIT Madras, Prof. P. J. Narayanan of IIIT Hyderabad, and Prof. Paul Scheunders of University of Antwerp, Belgium, for their comments.

Unlike an ordinary camera, hyperspectral imaging involves sophisticated instrumentation to capture the data. We are thankful to both JPL and USGS for making their AVIRIS and Hyperion data available for the researchers. Without such a data we could not have performed our experiments. A few of the figures used in this monograph have appeared in some of our publications. We are thankful to IEEE, Elsevier, and ACM for their copyright policy in allowing the authors to reuse these plots in their monograph.

The second author is thankful to the Bharti Centre for Communication, IIT Bombay for partial financial support in carrying out this initiative. The first author is thankful to DST for its support in the form of a J. C. Bose Fellowship. The administrative support from the publisher is also gratefully acknowledged.

Finally, the authors are thankful to their family members for their support and understanding.

Mumbai, February 2013

Subhasis Chaudhuri
Ketan Kotwal

Acknowledgments

Contents

Abbreviations

α	Fusion matte
β	Sensor selectivity factor
η	Additive noise
λ	Regularization parameter
Θ	Band selection threshold
$\mathbb{R}$	Space of real numbers
$\mathscr{E}$	Expectation operator
$\mathscr{F}$	Fusion rule
$\mathscr{P}$	Probability distribution function
$\mathscr{S}$	Computational Savings
μ	Lagrange multiplier
σ^2	Local variance
$G_{\sigma S}$	Gaussian spatial kernel
$G_{\sigma R}$	Gaussian range kernel
$\hat{b}$	Relative bias
C_i	i-th color channel
$\bar{g}$	Average gradient
$g\,(\bullet)$	Performance measure
F	Fused image
F_k	kth incrementally fused image
H	Entropy (average information)
I	Hyperspectral image
$\bar{\mathrm{I}}$	Subset of hyperspectral bands
I_k	kth image/band of hyperspectral image
k	Band index
K	Number of bands in hyperspectral image
L_i	ith norm
$m\,(\bullet)$	Mean value
$Q\,(\bullet)$	Quality factor
$\mathbf{s}(x,y)$	Spectral array at pixel (x,y)
w	Fusion weights
$X,\,Y$	Image dimensions

BD	Bhattacharyya Distance
FF	Fusion Factor
FS	Fusion Symmetry
HDR	High Dynamic Range
IHS	Intensity-Hue-Saturation
JSD	Jensen-Shannon Distance
LDR	Low Dynamic Range
MI	Mutual Information

Chapter 1
Introduction

The earth is full of different resources such as minerals, land, water, and vegetation. As majority of these vast resources have been proved to be a boon towards the progress of mankind, the identification of such resources has always been one of the primary goals. Earlier methods of discovery and identification of resources required a visit to the actual field followed by the collection of samples needed for the analysis of resources in the pure or mixed form. However, advances in technology have revolutionized the ways of collection, analysis, and management of data. Remote sensing has proved to be one of the highly effective technologies to observe our planet through a huge collection of data.

Remote sensing may involve acquisition of pictorial information about a particular scene or region on the earth without any kind of physical contact. Along with the natural resources, it also provides information about the man-made resources, e.g., urban land usage. The extensive information gathered is provided in a map-like format which makes remote sensing images a highly useful and a viable source. Additionally, it also facilitates a repetitive coverage of the same region on the earth, quite useful for understanding the temporal behavior. This vast information has served a large number of applications as well as research areas including resource exploration, geological surveying, mineral detection, environment monitoring, weather analysis, land cover, and vegetation analysis. Today, remote sensing has become an indispensable tool for the earth observation. The success of remote sensing can be attributed to many reasons. Firstly, it does not require a physical presence of the operator or the instrument at the site location. The images are captured by the sensors that are several thousand meters above the site. The satellites can easily collect the data from areas practically inaccessible to the man. Secondly, remote sensing technology has proved to be very economical. It covers large areas at a very low per unit cost. Also, the process of data collection is very fast. Lastly, with recent advancement in the remote sensing imagery, one is able to obtain more and more accurate and robust data.

One may regard the process of remote sensing as something that *reads* the data by means of various sensors. The sensors can obtain these data from different sources, where the most common sources include acoustic or electromagnetic waves. This

S. Chaudhuri and K. Kotwal, *Hyperspectral Image Fusion*,
DOI: 10.1007/978-1-4614-7470-8_1, © Springer Science+Business Media New York 2013

monograph discusses the remote sensing data obtained by the means of variations in the electromagnetic (EM) energy reflected from the surface of the earth.

Let us take a brief look at the process of EM remote sensing data acquisition. The sun is the primary source of electromagnetic energy which is propagated through the atmosphere. Elements of the earth's surface absorb some component of this incident energy which is partly emitted back, and reflect the remaining fraction. The amount of energy reflected back into the atmosphere primarily depends on the composition of the material, and thus varies from material to material. These variations in the reflected energy are captured by the remote sensing systems that are attached to either a satellite or an aircraft. This data is then digitized and converted into a pictorial representation which makes the process of interpretation and analysis easy. However, before this raw data could be analyzed by the subject experts of various application fields, it is necessary to convert it into a *suitable* form. The remote sensing information is often in the pictorial form, i.e., images. The intensity at each pixel corresponds to the digitized radiometric measurement of the corresponding surface area. In remote sensing community, the intensity value of a pixel is often referred to as the digital number. An ideal image is expected to accurately reproduce the reflectance of the surface measured within a particular wavelength band. The intensity value of a pixel is expected to be proportional to the product of the surface reflectance and the incident energy from the sun (spectral irradiance). However, in most of the cases, the acquired images suffer from various kinds of distortions. The atmospheric interference and sensor defects cause incorrect recording of the data in the sensor system, and thereby introduce radiometric distortion. The factors like the earth's curvature, and the perspective projection cause errors pertaining to the shape and size of the objects. These are referred to as the geometric distortions. Unless such distortions are corrected, these data cannot be effectively analyzed. Additionally, for different applications, different features of the data are required to be studied and highlighted. Enhancements of the remote sensing images also play a major role in the computer assisted analysis. The science of digital image processing has extensively been applied for the pre-processing of the remote sensing data. The subsequent analysis and decision making part comprises of the subject experts and analysts from various application fields. Figure 1.1 illustrates the step-wise schematic of the remote sensing process.

In this monograph, we discuss one of the important image processing applications that aims at providing a better visualization of the remote sensing data through combining multiple images of the same scene. These images provide complementary information about the scene which when integrated into a single image turn out to be a quick, efficient, and very useful visualization resource for the human analysts.

When a remote sensing system captures large areas, the altitude of the imaging system which could be an aircraft or a satellite, increases. Subsequently, this leads to the reduction in the size of various objects in the scene. The smallest element of an image, i.e., a single pixel may correspond to an area of a few squared meters in the actual scene, depending upon the instantaneous field of view (FOV) of the sensor. In a remote sensing image covering an area of a few hundred kilometers in both length and width, smaller objects such as buildings and roads occupy a very few pixels

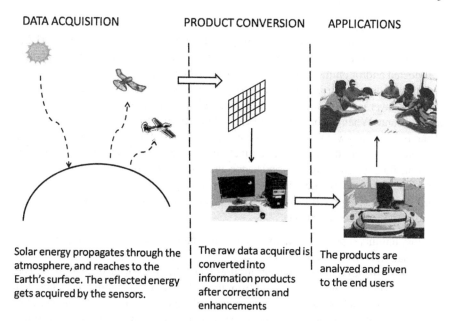

DATA ACQUISITION **PRODUCT CONVERSION** **APPLICATIONS**

Solar energy propagates through the atmosphere, and reaches to the Earth's surface. The reflected energy gets acquired by the sensors.

The raw data acquired is converted into information products after correction and enhancements

The products are analyzed and given to the end users

Fig. 1.1 Illustration of the remote sensing process

for their representation. Thus, the coverage of a wide area is compromised by the spatial resolution of images. The primary task of most of the remote sensing systems is to classify the captured region into various segments. These segments, typically known as land cover classes are decided by the application. Every pixel in the scene is associated with one of the classes. However, when the spatial resolution of an image is poor, two or more classes can be the part of a single pixel, and it may not be always possible to correctly extract the details. The design and operational cost of sensors with high spatial resolution is very high. Such sensors, when built, generate a huge amount of data in order to cover a wide geographical region. One would like to collect as much information as possible regarding the composition of the scene for a better classification. The reflectance of the materials present in the scene varies with the wavelength. Thus, the values of reflectance, if captured at different bands of wavelengths, provide disparate information related to the scene contents. This set of images can be obtained by collecting the response of the scene over multiple spectral bands. A cost effective solution that does not compromise with either the area coverage, or the spatial resolution, and yet provides a better information about the scene is built in the form of spectral imaging.

1.1 Spectral Imaging

The remote sensing data get captured by the sensor for a given spatial resolution at a particular wavelength. Various objects in the scene reflect, absorb, and emit electromagnetic radiation depending upon their material composition. For a single grayscale image, the airborne or the satellite sensor captures incoming radiation only at a given range of wavelength, and discards the rest of the electromagnetic (EM) spectrum. The discarded part of the EM spectrum contains variations in the amount of reflected and/or emitted radiation at those wavelengths.

If this incoming radiation is collected over a sufficiently narrow spectral band, one obtains an average radiance of the underlying material integrated over that particular spectral band. When the bandwidth of each of the spectral bands is sufficiently small, of the order of a few nanometers, one can practically obtain an *almost continuous* spectral response of the material as a function of wavelength. This response function is unique for any given material, and hence, in principle, can be used for identification purposes. This array of spectral observations is referred to as the *spectral signature* of the material. The spectral signature is one of the most important characteristics of the material for several reasons:

- Being unique, it provides unambiguous information towards the identification of the materials in the scene.
- A typical image captures the materials present on the outer surface of the earth. The spectral signature gets constituted from the radiance pattern of the materials that could be present beneath the surface of the earth, and thus not directly visible. This property makes the spectral signature particularly useful in the task of identifying the minerals beneath the surface of the earth.
- When there exists a mixture of two or more underlying materials within a single pixel, the captured spectral signature is comprised of the weighted sum of the spectral signatures of the individual materials. Using some spectral unmixing techniques, one can identify various materials present along with their proportions.

When the spectral response of the scene is captured by a single wideband sensor, the magnitude of the image pixel represents the integrated spectral response of the scene material. In this case, one obtains a single image of the scene—that is an integrated response of the scene over the bandwidth of the imaging sensor. Naturally, such images lack in their ability to discriminate among various materials. This shortcoming of the single sensor imaging gave rise to the idea of spectral imaging.

The idea of spectral imaging is based on the use of spectral variations in the electromagnetic field. Multispectral (MS) image captures the spectral information of the scene in the form of around 4–6 bands covering the visible, reflective infrared (IR), and thermal IR regions of the spectrum. Each of the bands represents radiation information acquired over a fairly wide-band in the wavelength spectrum. Thus, one can think of the multispectral data as the sampled version of the spectral response of the pixel. For most multispectral imaging systems, however, these bands are few, wide and separated; and therefore, the sampling of the spectral response is not dense. The multispectral (MS) images have several advantages over a single sensor image. The MS image captures spectral information from the bands that are beyond the visible range. This set of bands captures the reflectance response of the scene over the particular spectral bands, and thus, each band provides complementary information about the scene with respect to the other bands. With multiple observations of the scene element, the identification and classification capabilities of the MS data are

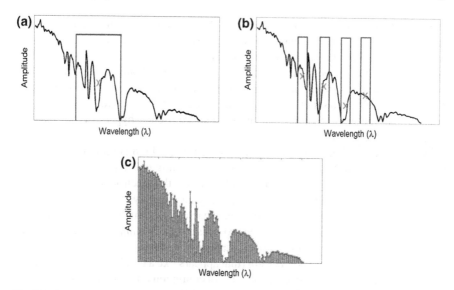

Fig. 1.2 The schematic representation of the spectral response of the scene captured by the different sensor systems: **a** the single sensor system, **b** the multispectral system, and **c** the hyperspectral system. The rectangle window defines the region of integration of the spectral response

better than those of the single sensor image. If the bands are captured corresponding to the wavelengths of the red, green and blue colors, then it is possible to obtain an RGB image of the scene along with other bands. Humans perceive colors better than the grayscale changes. As the MS data facilitates the color (RGB) visualization of the scene, human analysts and observers can have a better understanding of the scene. We illustrate the schematic comparison of how the data at a particular pixel would appear in the case of the MS image in Fig. 1.2. The rectangular block represents the response of an ideal single sensor. The output of this sensor is obtained by the integration of the reflectance response of the scene over the bandwidth of the sensor element. The entire response is represented by a single value which is the average value of the response over the sensor bandwidth. In the illustration, we represent this value by ($\times$). However, being a single number, it does not provide much information about the scene contents. On the other hand, the multispectral sensor captures the scene response at multiple wavelength bands. Figure 1.2b represents an illustration of the response by an ideal 4-band multispectral sensor. The scene response has been captured by 4 different sensor elements that have a lesser individual bandwidth than that of the single sensor as in Fig. 1.2a. Therefore, the responses of individual sensor elements capture the scene responses better. The multispectral bands, thus, are able to provide a more detailed information regarding the scene in the form of multiple data samples.

However, even the multispectral images do not meet the challenges posed by advanced applications of the remote sensing data. The small number of bands in the MS image is not enough to discriminate between two materials that are somewhat similar in their spectral responses. Additionally, the multispectral imaging sensors have a reasonably wide bandwidth. The pixel magnitude, therefore, is an averaged value of the response function over the corresponding spectral range which may not precisely reflect the spectral information. As the spectral bands in MS images are well-separated, the spectral information is not very dense and uniformly sampled. Thus, we do not have a set of contiguous and dense sampled spectral bands. In order to overcome these shortcomings of the multispectral data, hyperspectral imaging systems have been developed.

A hyperspectral imaging sensor acquires the reflected radiation in the form of a large number of (nominally over 50) narrow and contiguous bands. Hyperspectral system contains an array of sensors which samples the electromagnetic spectrum ranging from the visible to the near-infrared region typically covering wavelengths from 0.4 to 2.5 μm. The nominal bandwidth of each of the element of the sensor array is 10 nm. The hyperspectral (HS) image provides a densely sampled and almost continuous spectral signature over the given wavelengths. This is illustrated in Fig. 1.2c. Each narrow rectangle in this spectrum (that appears like a thin line in Fig. 1.2c) represents elements of the sensor which are almost contiguous. The hyperspectral image is, thus, a dense and uniformly sampled version of the continuous spectral response. This characteristic of high spectral resolution makes differentiation of various materials on the earth possible. The dense and contiguous nature of the data provides a very accurate and more complete information about the scene. Several materials radiate the incident light in a higher proportion only in a very narrow spectral range. Such objects appear prominent and observable only in these related spectral bands; while these objects could often be dominated by the other materials in their vicinity at all other wavelengths. Hyperspectral images provide an *almost continuous* sampling of the spectral data, and thus essentially capture even minor variations in the scene reflectance. This characteristic imparts tremendous advantages as far as the data classification is concerned. That is, it enables the user to *visualize* the scene at any given wavelength.

The set of hyperspectral bands together is also referred to as the hyperspectral data cube. The front face of this data cube provides a two dimensional, spatially sampled information, and the third dimension provides spectrally sampled information along the depth of data cube. Each band provides a narrow-band image of the field as *seen* by the sensor. Along the wavelength dimension, each pixel provides a spectral signature characterized by the materials at that location.

The main technical content presented in this monograph does not require the knowledge of the image acquisition systems, We shall provide brief information to the readers about some of the commonly discussed hyperspectral sensor systems. The airborne sensor systems are attached to the aircraft that flies over the terrain under investigation. The sensor elements scan a line across the ground in the direction perpendicular to that of the flight. An array of scan-line pixels constitutes the X-axis data of the hyperspectral image, while the data along the Y-axis gets generated due to the forward motion of the aircraft. The imaging system splits the incoming spectrum at every pixel into a large number of channels that are contiguous and narrow-band. This aforementioned process produces hyperspectral data with a fine spectral

Table 1.1 Specifications of AVIRIS system

Year of availability	1987
Number of bands	224
Spectral range	0.4–2.5 μm
Bandwidth at FWHM	9.4–16 nm
Spatial resolution	4–20 m
Scan line width	1.9–11 km

Source http://aviris.jpl.nasa.gov

resolution where the width of the HS image (X-axis) is determined by the sensor, and the length (Y-axis) depends on the distance covered by the aircraft. Although airborne hyperspectral systems have been operated since 80s, the significant milestone was the development of the Airborne Visible/Infrared Imaging Spectrometer (AVIRIS) designed by the Jet Propulsion Laboratory (JPL) of NASA [10]. The objective was to measure the upwelling spectral radiance from those areas of the earth where the solar reflected energy was dominant. The AVIRIS produced the first set of images in 1987 becoming the first earth-looking spectrometer covering the entire spectral range from 0.4 to 2.5 μm. The spectral resolution of hyperspectral systems is often reported in terms of full width at half maximum (FWHM) which is related to the width of an approximated spectral channel. For the AVIRIS sensor, this measure varies between 9.4 and 16 nm. This airborne sensor acquires the data in the form of 224 contiguous bands at 10 nm intervals. With 614 pixels per line and a spatial resolution of 20 m, it covers an 11 km wide region. These specifications of the AVIRIS have been summarized in Table 1.1. The AVIRIS applications range from geology, hydrology, environmental monitoring, as well as land and ocean ecology.

Some other airborne hyperspectral imaging systems include Compact Airborne Spectrographic Imager (CASI) by the Canadian company ITRES. The CASI has a programmable spectral band filters where one can obtain up to 288 spectral bands. CASI has been used for the GIS applications, and also for the vegetation analysis. Naval Research Laboratory developed the HYDICE for civic applications such as agriculture, geology, and ocean studies. The HYDICE, an acronym for the Hyperspectral Digital Imagery Collection Experiment, was one of the early airborne instruments with a high spatial resolution achieved by flying at a relatively low altitude. The HyVista corporation of Australia designed the HyMAP imaging spectrometer which covers the spectral range from 0.4 to 2.5 μm in 126 contiguous bands. The bandwidth of each channel is between 15 and 20 nm. The HyMAP is designed for the purpose of mineral exploration.

While airborne sensors have been in demand, the spaceborne sensors provide several advantages over the sensors from the former category, including:

- Coverage of wider areas on the earth.
- Repeated (multi-temporal) data collection.
- Reduction in distortion due to sensor motion.

Table 1.2 Specifications of Hyperion system

Year of availability	2002
Number of bands	220
Spectral range	0.4–2.5 μm
Bandwidth at FWHM	10 nm
Spatial resolution	30 m
Scan line width	7.5 km

Source http://eo1.gsfc.nasa.gov

The Hyperion was the first spaceborne hyperspectral sensor to acquire the reflectance from near IR and short wave infrared (SWIR) band [77]. The imagery is a part of EO-1 spacecraft launched by NASA. Hyperion provides 10-bit data covering a bandwidth of 0.4 to 2.5 μm in 220 bands. This instrument can cover a 7.5 km by 100 km region per image. Table 1.2 summarizes the basic specifications of the Hyperion.

The radiance acquired by the sensor cannot be directly used for further operations due to illumination and atmospheric effects. These effects mainly include the angle of the sun, viewing angle of the sensor, the changes in solar radiance due to atmospheric scattering [160, 165]. It also suffers from geometric distortions caused due to the earth's curvature. Further operations are made possible after atmospheric corrections which transform radiance spectra into reflectance. More details on the corrections can be found in [29, 160, 168]. The raw imagery obtained from the sensors is referred to as the Level 0 data. At the next level of processing, called Level 1R, the data are calibrated to physical radiometric units, and formatted into image bands. Later, the data are geometrically corrected, compensated for atmospheric effects, and localized on the earth. These data are referred to as the Level 1G. One of the important requirements for the processing and better understanding of these data is to have all the bands of the hyperspectral image depicting exactly the same region. In other words, all the bands should be spatially aligned to each other. This process is known as image registration. We deal with only the co-georegistered Level 1G data in this monograph.

Figure 1.3 illustrates the 3-D nature of the hyperspectral data. It is often convenient to visualize the hyperspectral image as a 3-D cube. One may observe a set of 2-D bands that can be thought as the "slicing" of the cube along the spectral axis. When the spectral signature is to be analyzed, it can be accomplished by observing an 1-D array of intensity values across all the bands for the specific spatial location under investigation. The left portion of the figure depicts one of the bands of the hyperspectral data. Observations along the wavelength axis at any spatial location form an array of length equal to the number of bands. Two such representative arrays are shown in the plot. These plots refer to the spectral signatures of the material compositions of the corresponding pixels.

The 3-dimensional structure of the hyperspectral images indicates three different structures for the storage of the data. If the hyperspectral image is to be primarily regarded as the set of independent bands, it is beneficial to store it band-wise.

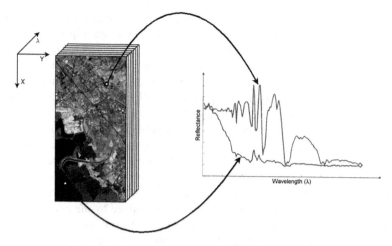

Fig. 1.3 Schematic representation of the 3-D hyperspectral data cube and illustration of the spectral signature

The first band of the hyperspectral data is stored as a stand alone image. A 2-D image array is stored by collating the rows—which is a standard format for saving a 2-D data. This format is known as the band sequential (BSQ). If the hyperspectral image is to be used for classification related activities, one needs to work with the spectral signatures. In such a case, it is desirable to store the data as a contiguous collection of 1-D spectral arrays. The band interleaved by pixel (BIP) structure stores the hyperspectral image as a set of location-wise arrays for all the spatial locations in the data. The BIP format facilitates faster reading and processing when individual spectral signatures are to be dealt with. However, in order to observe any of the bands, one needs to read the entire data, extract the necessary pixel values, and rearrange them. A compromise between these two storage formats is provided by the Band interleaved by line (BIL) format. In the BIL format, one stores the intensity values of the first scanline of the first band, followed by the fist scanline of the second band, and so on. After storing the first scanline values for all the bands in the data, one proceeds towards the second scanline and the process continues till the entire data gets stored. Table 1.3 summarizes the three formats. If the hyperspectral data contain K bands of dimensions $(X \times Y)$ pixels each, the structures for each storage format can be described as given in Table 1.3.

Table 1.3 Storage formats for hyperspectral data

Storage format	Acronym	Data structure
Band sequential	BSQ	$X \times Y \times K$
Band interleaved by pixel	BIP	$K \times X \times Y$
Band interleaved by line	BIL	$X \times K \times Y$

Hyperspectral imaging has been found to be highly useful for a wide span of application areas. Every object on the earth's surface possesses a specific pattern of energy reflection across the electromagnetic spectrum. Hyperspectral data have two unique characteristics—a very fine spectral resolution, and a large number of bands. This advancement of imaging sensors and the growing power of computers have enabled researchers to explore various applications of hyperspectral data. During the initial phase of research, the hyperspectral data were mainly used for surveillance purposes by the military. These images reveal spectral characteristics of the objects which can potentially be used to uniquely identify them. Due to their classification ability, hyperspectral images have been used in defense and military to detect various camouflages. It could be used to detect hidden objects where employing a conventional RGB camera fails due to its poor spectral resolution. It is also possible to make use of hyperspectral imagery to detect a camouflage by the enemy from the vegetation or the soil. Automatic spectral target recognition is an important military application of hyperspectral data. Traditional methods of target detection are based on the spatial recognition algorithms. With the advent of hyperspectral imagery, it has become possible to reveal the spectral characteristics of the target resulting in a better target identification [152]. The HYMEX project of the Canadian defense has been intended towards detection of military vehicles, camouflages, and man-made materials [6].

Some of the bands in hyperspectral data are affected by various atmospheric constituents. These include water vapour, carbon di-oxide, aerosols, etc. The analysis of these bands brings up the information about the corresponding factors affecting the band. For example, the analysis of the bands near 0.94 μm is useful for the detection of the presence of water vapour [153].

Hyperspectral data provide an accurate information about the scene and its composition. It has proved to be a rich source for various geo-exploration activities. The remote sensing images can be used to explore minerals, soil, snow, etc. Minerals are usually found beneath the earth's surface in a pure or a mixed form. The sensors AVIRIS and Hyperion have demonstrated the ability to remotely map the basic surface mineralogy [93], while the data obtained from the HyMAP hyperspectral sensor have also been analyzed for the purpose of mineral exploration along with accurate map generation, and soil characterization [74].

The hyperspectral data are capable of revealing various characteristics of the soil such as its type, moisture content, and erosion. With the help of readymade soil spectral libraries, it is possible to determine the material decomposition of the soil [6]. The fertility level of the soil and its variation over a spatial neighborhood within a field can be investigated with the aid of hyperspectral data [75].

The list of application areas is continuously growing. Typical applications include environment monitoring [157, 170], characterization of land and ocean areas [42, 129], characterization of ice and snow areas [53, 121, 154], crop and vegetation assessment [38, 117], etc. Hyperspectral data have also proved to be quite useful for medical applications [56, 161, 179], and industrial applications [69, 171]. Readers are requested to go through these references for detailed information. The focus of this monograph is primarily into remote sensing applications. Therefore, our illustrations are based on the remote sensing images.

As explained above, the analysis and applications of hyperspectral data have been explored in various areas, and they have been proved to be highly effective. However, the data analysis imposes several challenges:

1. A hyperspectral image contains nearly 200–250 bands. For an airborne sensor, the dimensions of the images are governed by the scanline width and the time of flight. The field of view (FOV) along with the altitude covers a large terrain on the earth in case of a spaceborne sensor. Storage of such data requires terrabytes of memory.
2. The computational costs of data processing are proportional to the size of the data. Due to high volume of the hyperspectral data, the cost of processing is also very high.
3. The spectral response of a scene does not change drastically over the adjacent bands. These bands exhibit a very high degree of spatial correlation due to the contiguous nature of the hyperspectral sensor array. Thus, hyperspectral data contain a high amount of redundancy.

The standard display systems are tristimulus in operation. That is, the display systems can accept three different images as inputs, and assign them to red, green, and blue channels to generate a color (RGB) image. Alternatively it can accept only a single image to produce a grayscale image output. The hyperspectral data contain a far more number of bands than those can be displayed on a standard tristimulus display. The appearance of the objects is based on their spectral response. Certain objects in the scene appear prominent in some bands for which their constituent materials exhibit a large reflectance, while they can practically disappear in some other bands due to a very poor value of reflectance. Therefore, an observation of merely 1 or 2 bands does not provide a complete information about the data content. One has to go through all 200+ bands to understand the contents of the image. Also, in order to understand the spatial relations among two or more objects that are prominent in different spectral ranges, the observer needs to manually process the different bands so that the spatial alignment of features across them can be well understood. To address the aforementioned associated problems and to exploit this rich source of data, an efficient technique for visualization of hyperspectral image could prove to be a primary and important processing step. If a human analyst is able to quickly observe the contents of the scene, s/he can initiate the necessary data-specific image processing algorithms. One can save a significant amount of computation by choosing a set of appropriate processing algorithms once the data contents are roughly known. Visualization of hyperspectral image is thus, very useful preprocessing task for most applications. This visualization can be accomplished by means of fusion of bands of hyperspectral image which is a process of combining data from multiple sensors. Here, the goal of fusion is to form a single image that captures most of the information from the constituent bands of the data. Image fusion is a vast area of research spanning fields from computational photography to medical imaging.

Image fusion is the specific case of the sensor data fusion where all the sensors refer to the imaging devices. Image fusion is a tool used in multiple disciplines of technology to maximize the information from a set of input images for a particular application. It is defined as the process of combining information from two or more images of a scene into a single composite image that is more informative and is more suitable for visual perception or computer processing [68].

1.2 Hyperspectral Image Visualization

Consider a hyperspectral data cube consisting of nearly 200 bands. For visualization purposes, we want to represent most of the information in these bands using a single (grayscale or color) image. We want to create a single image through a process known as image fusion, that preserves as many features as possible from the constituent bands for the purpose of visualization by a human observer. If any specific application requires a certain set of features to be highlighted, one may accordingly select and enhance only that subset of hyperspectral bands where these features are clear and prominent. However, this knowledge regarding the application or the imaging system limits the scope of the fusion process, and makes it completely application dependent, and thus, it cannot be used over a large class of data as a generalized process. We do not assume any knowledge regarding the parameters of the hyperspectral sensor. Thus, the fusion system is expected to be designed in a blind way which generates the output from the hyperspectral data only, for any general class of applications.

Fusion can be performed to accomplish various objectives. These objectives and applications are primarily responsible for the choice of fusion methodology. Fusion has often been used as a means for providing a quick visualization to humans [20]. Multiple images capture different aspects of the scene. To visualize this disparate information, the observer has to view all the images independently, and then manually combine the different pieces of data across the same spatial location provided by the constituent images. One can obtain a quick visualization of the scene using the reduced dataset through fusion. An appropriate fusion scheme generates a single (grayscale or color) image which preserves most of the features of the input images that can be considered to be useful for human visualization. A fusion is sometimes considered to be a step prior to the classification. The remote sensing data is often used for mapping of resources such as minerals, vegetation, or land. This mapping involves classification of the scene pixels into a set of pre-defined classes. The multiple bands provide complementary characteristics of the data useful for the identification of the underlying materials. We have already explained in the previous section the usefulness of the hyperspectral images in terms of providing a dense sampling of the spectral signature of the pixel. However, due to a very large number of bands, the classification process turns out to be computationally highly expensive. One would like to retain only those data that contribute towards a better classification, and discard the remaining redundant part. A classification-oriented fusion provides an effective solution to this problem. Here the main objective of fusion is to combine the information across multiple bands in order to increase the classification accuracy of the so obtained resultant image. We often come across images of the same scene with different spatial resolution. The spatial resolution refers to the size of the smallest discriminable object in the image. Images with higher spatial resolution are desirable, however they may not be always available due to high cost involved in such imaging sensor design. The issue of improving the spatial resolution is often solved through fusion of low-resolution images with the high-resolution image of the same scene. In remote sensing, this fusion is also known as the pan-sharpening which explores the sharpening objective of fusion. We provide more details on pan-sharpening in Chap. 2. The focus of the monograph lies on fusion of hyperspectral data purely for the purpose of visualization intended for a human observer.

As we traverse along the spectral dimension, different values of pixel intensity can be observed at the same spatial but different spectral locations. We need to incorporate these values into the fused image according to their relative importance. Images with a high value of contrast and sharp features appear visually pleasing. However, images with a large number of over- and under-saturated pixels do not possess much visually informative content. As visualization is the focus of this monograph, we want the fused image to have these desirable characteristics. While satisfying these requirements, the fusion procedure should not introduce any visible artifacts. The problem of visualization of hyperspectral data over a standard display device is thus, quite a challenging problem as hundreds of bands together encompass a large volume of disparate information.

As the theory of image fusion began to develop, it was highly important to measure the performance of such fusion systems. This problem becomes challenging when the reference image or the ground truth is not available. Researchers have proposed a few measures based on the statistical information such as entropy of the image. Most of these measures have been defined for a generalized fusion of images where the number of images being fused is very less, say 2–6, and thus, the calculations of statistical performance measures are easily implementable. The problem of evaluation of fusion is more difficult in the case of hyperspectral image fusion due to a large number of bands, high volume of independent information among them, and unavailability of any reference. The existing measures of image fusion should be modified specifically for the fusion of a very large number of bands. We shall explain several adaptations of existing measures to analyze the fusion techniques for hyperspectral images in a better manner.

1.3 Tour of the Book

This monograph addresses the problem of fusion of hyperspectral image bands for the purpose of visualization. We do not assume any knowledge of hyperspectral sensor parameters and scene contents. Having obtained the fused images, the problem of performance evaluation of the fusion techniques is quite challenging, particularly due to a large number of constituent bands, and unavailability of any reference image. We also address the problem of evaluation of different techniques for fusion of hyperspectral data. The highlights of the book are discussed below.

- First, we provide an overview of the process of image fusion. We familiarize the readers with various concepts related to image fusion. Then we discuss several techniques of generalized image fusion, and specific techniques of hyperspectral image fusion. We also brief the readers with some of the performance measures for quality assessment of fusion techniques.
- For an effective visualization of the entire hyperspectral data cube, it is naturally expected that the fused image should retain as many features of the data as possible. One would especially like to preserve the weak features of the data during the process of fusion for a better visualization of the scene. Most of the fusion techniques generate the resultant image as a weighted linear combination of the set of input hyperspectral bands where the choice of weights is the most critical aspect. We discuss a methodology that assigns appropriate weightage to the weak textures in the data so that these can be prominently represented in the fused image. We explain the use of an edge-preserving filter known as bilateral filter to extract the textural content of the hyperspectral data. We define the fusion weight for every individual pixel based on the amount of textural content at that location obtained as the difference between the original pixel and the output of the bilateral filtering at the same location. Through this weighing scheme, we assign higher weights to the weak edges and features in the data that exist over a very few bands as they might get lost in the process of fusion otherwise.
  Fusion of hyperspectral data involves processing over nearly 200+ bands. For processing all the bands together, one is required to read the entire hyperspectral data cube into the memory. Due to a high volume of the data, the memory requirement often goes beyond a few hundreds of megabytes. Furthermore, when a large number of bands are being used together to assign weights, some pixels might get assigned with very small values even comparable to the truncation limits of the storage system. This leads to a risk of loosing the contribution of some of the pixels towards the final result. We discuss a hierarchical scheme of fusion to prevent the aforementioned problems of hyperspectral image fusion. In this scheme, we split the hyperspectral data into several subsets where each of the subset contains nearly 10 % of the original number of bands. We fuse these subsets independently using the bilateral filtering-based solution. Since only a smaller chunk of the data is required, the problems of memory requirement and smaller weights are circumvented in this scheme. Also, the fusion of each subset is independent of other, which provides a scope for possible parallelization in order to speed up the entire process.
- Most pixel-based fusion techniques compute the fusion weight for every pixel in every band of the input hyperspectral image. The adjacent bands in hyperspectral image exhibit a very high degree of spatial correlation as they depict the reflectance response of the scene over contiguous wavelength bands. The successive bands, thus, contribute a very little additional information towards the fusion process. Therefore, the addition of such bands brings a nominal information gain to the

fused image, however, at the cost of computation of the fusion weights which might be computationally expensive, especially when employed on a per pixel basis. We develop an information theoretic strategy for selection of only a few, but specific hyperspectral bands, which capture most of the information content in the hyperspectral data. We select a subset of hyperspectral bands that are mutually less correlated with each other in order to minimize the redundancy in the input hyperspectral data. A particular hyperspectral band is selected for fusion only when it contains a significant amount of additional information as compared to the previously selected bands. The band selection scheme is independent of the fusion technique to be employed, and thus, the subset of selected bands can be fused using any pixel-based fusion technique. The hyperspectral bands are typically ordered as per their spectral wavelengths. We develop a model for the conditional entropy of the bands as a function of the spectral distance between them. We also discuss a special case of band selection for this spectrally ordered data. This scheme provides a computationally efficient and fast way for the selection of subset of bands. An appropriate combination of these bands provide a fused image with a minimal loss in the visual quality as compared to the output image formed through fusion of the entire hyperspectral data using the same fusion technique. We also provide theoretical bounds on how much one can save in computation as a function of number of bands selected.

- An array of observations at a given spatial location is known as the spectral signature of the corresponding pixel. The reflectance response of the scene elements being sensitive to the wavelength, some of the scene regions get captured with high values of intensity in only a fraction of total number of image bands. Therefore, it may be observed that not every pixel carries an equal amount of information required for the visualization-oriented fusion. The pixels that provide visually important information towards fusion should contribute more. We employ a model of image formation which relates the input hyperspectral data and the resultant fused image through a set of parameters referred to as the sensor selectivity factor that quantifies how well the scene has been captured by the corresponding sensor elements. We determine the visual quality of the pixel without the availability of any ground truth. We develop a strategy to compute these parameters of the image formation model using some of the qualities of the data that relate to the visual quality. In this technique, we consider the well-exposedness and the sharpness to be the quality measures along with the constraint of intra-band spatial smoothness in model parameters.

In order to estimate the fused image, i.e., the true scene, we employ a Bayesian framework. However, instead of using the L_2 norm-based priors which tend to reduce the sharpness of the image, we incorporate a total variation (TV) norm based prior which is based on the L_1 norm of the image gradient which brings smoothness to the resultant image, and yet preserves the sharp discontinuities, and thus the edges. The solution provides fused images that are sharp and visually appealing. This solution provides a two-fold flexibility. First, one may choose from a variety of quality measures to efficiently capture the pixels important for fusion. Secondly, the problem of fusion is posed as a statistical estimation problem for which a large number of statistical tools are available. Thus, one may experiment with a variety of estimators, priors, and implementations.

- First two fusion techniques discussed in this monograph explicitly compute the fusion weights from the input data. These weights are also known as fusion matte, or more commonly α-matte in the graphics literature. However, an explicit presence of a matte is not required if the fusion model is well-specified. In this technique, we model the weighting function as an input data-dependent term, and provide an iterative solution that generates the fused image without explicit computation of the fusion weights (or mattes). The weighting function has been derived from two important aspects of fusion—first, we want fusion weights to be proportional to the local contrast of the pixel which we calculate as the local variance. We also expect the intensity of the fused pixel at any location to remain close to the intensities of all the constituent pixels from input bands at the given location. This second aspect tries to minimize the radiometric distortion in the fused image. Another important requirement for visualization-oriented fusion is the natural appearance of the fused image with no visible artifacts. As most of the natural images are spatially smooth, several smoothness-based constraints are often employed in various problems of image processing. We incorporate the smoothness constraint on the fused image that penalizes the discontinuities in the image, and formulate the problem in the framework of variational calculus. The solution is provided using the Euler-Lagrange equation which iteratively refines the fused image for the above mentioned objectives. The final fused image is formed by an appropriate combination of pixels with higher local variance, and at the same time it minimizes the radiometric distortion in the fused image with respect to the input hyperspectral data. As there is no explicit computation of fusion mattes, one may refer to this as a matte-less approach to hyperspectral image fusion.

- Most of the fusion techniques including these discussed above define the fusion weights from certain properties of the data. Therefore, the properties of the input define how the data should be combined, and thus drive the fusion process. The goal of fusion is related to an efficient visualization of the scene by a human observer. The fused image is, thus, expected to have certain properties that are considered to be desirable for a better visualization. For example, well-exposed and high local contrast in the fused image. Given the input data, the existing fusion techniques do not guarantee to what extent such properties will be satisfied in the resultant image. Interestingly, one can think of the fusion problem from a completely different perspective. One may aim at the generation of a fused image with such desired properties, irrespective of the characteristics of the data contents. We formulate a multi-objective cost function based on these properties and transform fusion into an optimization problem. Likewise in the earlier technique, we consider the well-exposedness and contrast on a per pixel basis. In order to obtain a regularized solution, one may enforce a smoothness constraint on the output image. However, this constraint often leads to over-smoothening of the fused image, blurring of edges, and reduction in the contrast. In order to acknowledge the spatial correlation without degrading the contrast in the resultant image, we enforce the smoothness constraint over the fusion weights, rather than the fused image.

The fused image is generated from a combination of these spatially similar weights which preserves the discontinuities in the data. The fused image, thus, possesses very high values of contrast and sharpness without driving the individual pixels into over- or under-saturation. By using an appropriate auxiliary variable, we show how the constrained optimization problem can be converted to a computationally efficient unconstrained one.

- We have already discussed the band selection scheme based on the additional information or the conditional entropy of the successive hyperspectral bands. This scheme exploits the redundancy in the hyperspectral bands in order to select a specific subset of mutually less correlated bands. Consider the optimization-based fusion technique which generates a fused image that possesses certain desired characteristics. While we have explored an output-driven fusion method, one may as well want to investigate a band selection strategy based on the output fused image. We explore the idea of the selection of the specific hyperspectral bands based on the fusion output. This band selection scheme selects a band for fusion if the given band possesses significant additional information as compared to the fused image obtained by combining all bands selected so far. In other words, a given band is selected for fusion when the conditional entropy of the band with respect to the corresponding intermediate fused image exceeds a certain threshold. Naturally, this band selection scheme is specific to the fusion technique to be employed, which also governs its performance. In this scheme, the conditional entropy of the band is evaluated against the output image, rather than input bands. Therefore, if the band is visually quite different from the present output, it get included in the subset that produces the subsequent fused image. We can, therefore, obtain the fused image which captures most of the independent information across the bands using a quite less number of bands. The objective assessment of the fused image, however, shows a very minimal degradation in the quality of the image as compared to the image obtained from fusion of entire hyperspectral data using the same fusion technique.
- While the techniques for fusion of hyperspectral images are being developed, it is also important to establish a framework for an objective assessment of such fusion techniques. Such an assessment provides uniformity in the process of evaluation across various fusion techniques, and is also useful to explore various salient characteristics of the fusion technique. The extension of existing measures towards the evaluation of fusion of hyperspectral images is a non-trivial task. This problem is difficult due to a very large number of input bands, and absence of any reference image or the ground truth. We explain how one can extend some of the existing measures for evaluation of hyperspectral image fusion techniques. We also discuss several modifications in some of the measures for a better and efficient analysis of the fusion process. As a large number of techniques for generalized image fusion have been developed, one would like to extend these for the fusion of a very large number of images such as the hyperspectral data. We explain the notion of *consistency* of a fusion technique as more and more images are being fused using the same fusion technique. This consistency analysis can be very useful in

deciding suitability and adaptability of any technique towards fusion of a large number of images.

It is also important to understand how each of the bands contribute towards the final result. We provide several modifications to analyze the fusion techniques based on how the constituent bands participate in the process of fusion.

- We compare the performances of the fusion solutions discussed in this monograph (Chaps. 3, 5–7) in Chap. 10 using some of the measures discussed in Chap. 9. This chapter provides an extensive performance evaluation of all these techniques using several datasets. We also compare the performances of these techniques with some of the commonly adopted techniques for visualization of hyperspectral data.
- We draw conclusions from our work, and also provide possible future extension for the same in Chap. 11.

Chapter 2
Current State of the Art

This chapter presents past research efforts in the subject area related to this monograph. An extensive study of the existing literature which forms the guidelines for the development of new methodologies discussed in the monograph has been performed. We begin with the definition of the process of fusion and discuss some of the nomenclature in the field. We familiarize the reader with an overview of various techniques of fusion. The purpose of this chapter is to make the readers aware of various methodologies used for fusion, although not necessarily for remote sensing applications alone. Then we explain various existing solutions for visualization of hyperspectral data. With many techniques being developed, the problem of quantitative assessment of them is also gaining importance. This chapter also explores the existing measures for an objective assessment of fusion techniques. An overview of some of the evaluation techniques for fusion of hyperspectral data has also been provided. At the end of this chapter, we describe the notations used in the subsequent chapters of this monograph.

2.1 Introduction

During the past century, we have witnessed a rapid growth in the advancement of technology. Invention of a large number of sensors has resulted in high volumes of data collection. These datasets, obtained from multiple and multimodal sensors provide complementary information about the scene or objects being imaged. The performance of many intelligent systems and devices has significantly improved with the availability of the multi-sensor input. However, instead of processing individual inputs from all the sensors, it would be desirable to extract and merge the useful information from the set of sensors, and process this selective information in order to achieve a higher throughput. The usefulness of the information is, of course, dependent on the final application it is being used for. Therefore, to use the high amount of data in an effective manner, and to obtain a better performance from the

S. Chaudhuri and K. Kotwal, *Hyperspectral Image Fusion*,
DOI: 10.1007/978-1-4614-7470-8_2, © Springer Science+Business Media New York 2013

analytical system, one would like to combine the multiple sources of information together in such a way that the final representation contains a higher amount of useful information than any single input source. The process of combining or merging the data in a selective manner is known as *data fusion*.

Multisensor fusion has been defined as the synergistic combination of different sources of sensory information into a single representational format [203]. The aim of multisensor fusion is to utilize the information obtained from a number of sensors to achieve a better representation of the situation than which would have been possible by using any of the sensors individually. Image fusion represents a specific case of multisensor data where all the sensors are imaging sensors. Image fusion can be considered as a subset of a much wider field of data fusion and it has been receiving a large attention since the past 25 years. The need for fusion of images arose from the availability of multisensor data in various fields such as remote sensing, military applications, medical imaging, and machine vision. The process of image fusion refers to the combination of several images depicting the same object, but each of the images enhancing some particular features of the object. The primary objective of fusing images is to create an image that is more informative and more relevant to the particular application. The field of image fusion being very broad, it has been defined in several contexts. A generalized definition of image fusion can be stated as—*"the process of combining information from two or more images of a scene into a single composite image that is more informative and is more suitable for visual perception or computer processing"* [68].

Objectives of fusion differ with the applications. A single imaging sensor is often unable to provide a complete information of the scene. The process of fusion aims at integrating the complementary information provided by the different sensors for a better interpretation of the scene. Since early days, fusion has been mainly considered as a means for presenting images to humans [20]. An initial and the most direct technique of fusion was to sum and average the constituent input images. However, as authors in [20] have pointed out, this technique does not produce satisfactory results. Consider certain features that are prominent in only one of the inputs images. When averaging-based fusion is applied, such features are rendered with a reduced contrast, or they suffer from the superimposition of the other features from the remaining input images. A sharp image is much easier for humans to perceive. It is also desirable in several machine vision applications. Thus, fusion has been considered as one of the tools for sharpening of images [33]. The sharpening may also involve an increase in the spatial resolution of the image. This is one of the widely pursued objectives of image fusion, especially in the remote sensing community. A high resolution pan-chromatic image is fused with the low resolution data typically in the form of multispectral images. This process, known as the pan-sharpening, is a standard technique employed for enhancing the spatial resolution of the multispectral data. The multispectral data itself might be available in the form of a color (RGB) and a thermal image pair. Due to difference in the physics of the sensors, the visible and thermal images provide complementary views of the scene. A simple surveillance system incorporating an RGB camera along with the thermal camera can be used to identify security threats in public places. Fusion of the data streams obtained from these two complementary sources enhance the semantic capability of the surveillance systems.

The medical community has greatly benefited from the feature enhancement characteristics of image fusion. Medical diagnosis can be improved by the use of complementary information provided by the multimodal images such as computed tomography (CT), magnetic resonance imaging (MRI), and Positron emission tomography (PET). Fusion helps in enhancing features which are impossible to detect from a single image, and thus improves the reliability of the decisions based on the composite data [45].

Another objective of fusion is to remove the uncertainty through redundancy. Multiple sensors provide redundant information about the scene, however with different fidelity. The fusion of redundant information reduces the overall uncertainty, and leads to a compact representation. Thus, fusion can also be explored towards decision validation.

Through image fusion, we seek to obtain an effective way of integrating the visual information provided by different sensors. The redundant data should be removed without any significant or visible loss to the useful spatial details. In a routine case, the human observer has to go through the sequence of images, or needs to view all the images simultaneously in order to fully comprehend the features of the scene, and to understand their spatial correlation within the image and across the sequence of images. Fusion systems enable one to consider only a single image that preserves most of the spatial characteristics of the input. Having to consider only one displayed image at any one time significantly reduces the workload of the operator. Also, a human analyst cannot reliably combine the information from a large number of images by seeing them independently. An enhanced visualization of the scene with more accurate and reliable information is also one of the important objectives of image fusion. Furthermore, fusion systems considerably reduce the computation time and the storage volume for further processing.

2.1.1 Classification of Fusion Techniques

Since a large number of fusion techniques have already been developed, the classification of techniques helps in understanding the concepts related to fusion in a better manner. One may classify the fusion techniques in various ways. We discuss some of the fusion categories here.

Based on domain: A fusion algorithm can operate over the spatial data, i.e., directly over the pixel intensities to produce the final fusion result in the spatial domain itself. Alternatively, using different transforms such as Fourier transform, one may transform the set of input images into frequency domain. The fusion algorithm processes the frequency domain data to produce the result of fusion in the frequency domain. This result requires a reverse transformation such as inverse Fourier transform to obtain the fused image.

Based on resolution: Images to be fused may not have the same spatial resolution. This case is more frequent in remote sensing community. When the spatial resolution of the images is different, one is essentially enhancing the details in the images with lower spatial resolution using the remaining images. This process is also referred to as the pan-sharpening. In other cases, images have the same spatial resolution. These multiband images could be obtained from different sensors, at different time, or even at different spectral wavelength bands.

Based on nature of images: This categorization is somewhat different from the previous ones. Here one is more concerned about the type of the data rather than type or technique of fusion. The sources of images can be very different. Most of the real world scenes encompass a very high dynamic range (HDR). Most of the digital cameras are not able to capture these scenes due to their limited dynamic range. However, one can capture multiple images of the scene with varying exposure settings of the camera. This set of low dynamic range (LDR) images when appropriately fused, generates a single image that provides an *HDR-like* appearance [89, 145]. Such type of fusion is often regarded as multi-exposure image fusion. Similarly, the finite size of the aperture of the camera leads to defocused objects in the image. Due to the physics behind the camera lens, only the regions at a certain distance from the focal plane of the camera can be captured in focus for a given setting of the camera focus. To obtain a single image where all objects are in focus, we may capture multiple images by suitably varying the focus of the camera, and fuse them later. This multi-focus image fusion operates on different principles than those of multi-exposure images due to the difference in the formation of these images. In remote sensing, one often comes across multispectral image fusion where typically 4–10 bands of a multispectral image are combined to yield a compact description of the scene. Advanced hyperspectral imaging sensors capture the scene information in hundreds of bands depicting the spectral response of the constituent materials of the scene. Hyperspectral image fusion refers to the combining of these bands into a single image that retains most of the features from input hyperspectral bands. In the subsequent chapters of the monograph, we shall explore different techniques of hyperspectral image fusion. The medical community makes use of images obtained from different sensors (e.g., Positron emission, X-rays) providing complementary information about the scene. The fused image combines and enhances features from the input images which has proved to be quite useful in medical diagnosis. This class of fusion is referred to as multi-modal image fusion.

Based on processing level: One may consider the image as a two dimensional array of individual pixels, or as a set of certain features relevant to the application of interest. Alternatively, one may look at the image as the means of providing a decision, such as diagnosis of a disease, presence of security threatening objects, or existence of water bodies within some areas. Accordingly, images can be fused at pixel-level, feature-level, or decision-level. We can categorize fusion techniques with respect to the level of processing at which the actual fusion takes place. This categorization is particularly important because it decides the level of image representation where the actual fusion takes place. A similar categorization can be found for generalized data fusion as well. Let us take a closer look at fusion categories depending on the level of representation.

1. Pixel (or signal) level,
2. Feature (or region) level, and
3. Decision (or symbolic) level.

A general data can be analyzed and fused at the signal level which is the most basic and fundamental level of understanding and processing the data. The pixel level, or the pixel-based fusion regarded as the counterpart of signal level operations in the field of data fusion, is the lowest level of image fusion. Images from multiple sensors capture their observations in the form of pixels which are then combined to produce a single output image. Thus, the pixel-level fusion algorithms operate directly over raw data.

Feature level fusion requires feature extraction from input images with the use of advanced image processing operations such as region characterization, segmentation, and morphological operations to locate the features of interest. The choice of features plays an important role here, which is primarily decided by the end application. The regions or features are represented using one or more sets of descriptors. Multiple sets of such descriptors provide complementary information, which are then combined to form a composite set of features. These techniques are less sensitive to pixel-level noise [135].

For a decision level fusion, the input images and/or feature vectors are subjected to a classification system which assigns each detected object to a particular class (known as the decision). The classifier systems associate objects to the particular class from a set of pre-defined classes. Decision fusion combines the available information for maximizing the probability of correct classification of the objects in the scene which is achieved using statistical tools such as Bayesian inference. Figure 2.1 illustrates the relationship among all three levels of fusion according to their processing hierarchy.

Regardless of the fusion categorization, most of these fusion techniques require the set of input images to be spatially aligned, i.e., they represent exactly the same scene. This process of aligning is referred to as image registration which is quite a mature research field by itself. The discussion on different methods of image registration is beyond the scope of the monograph. However, throughout this monograph, the hyperspectral data is assumed to be co-georegistered, i.e., the spectral bands of the data depict exactly the same area, which is a quite common assumption in image fusion community.

2.1.2 Pixel-Level Fusion Techniques

This monograph discusses pixel-based fusion techniques where the final result of fusion is intended for human observation. As stated before, pixel-level fusion tech-

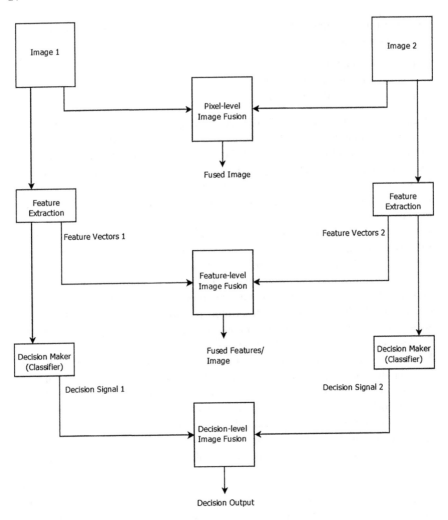

Fig. 2.1 Illustration of different levels of image fusion

niques can further be distinguished, depending on whether the images are fused in the spatial domain, or in the transform domain. The pixel-level fusion in the spatial domain is considered as a local operation where the individual pixels are processed to form the pixels of the fused image. The transform domain techniques, however, generate a fused image globally. The change in a single coefficient of the transformed image can reflect into a global (image-level) change in the output. This may lead to some undesirable effects in the fused image [205].

A large number of pixel-based techniques which span a whole plethora of applications have been proposed over the years. Some of these techniques have proved to be very successful even across multiple application domains. The scope of this monograph is limited to remote sensing applications, more specifically hyperspectral images. However, we also provide a brief description of fusion methodologies pertaining to the other application areas.

Remote sensing has been one of the leading image fusion applications with a large number of dedicated publications. The research in this area has been continuously growing as more precise and sophisticated imaging devices are being used. In 1999, Pohl and van Genderen have presented an in depth review of the existing work in multisensor image fusion till then [140]. This work covers a comprehensive range of fusion techniques with their objectives. It covers basic arithmetic techniques such as addition or ratio images, computationally demanding subspace-based techniques based on the principal component analysis (PCA), and wavelet-based multi-resolution techniques as well. This article introduces a number of applications of fusion including topographic mapping, land usage, flood monitoring, and geology. Furthermore, some pre-processing techniques and commonly used fusion schemes have also been reviewed.

Since a number of groups are actively working in the area of image fusion, the meaning and taxonomy of the terms differ from one to another. The establishment of common terms of reference helps the scientific community to express their ideas using the same words to the industry and the other collaborative communities. Wald presented a report on the work of establishment of such a lexicon for data fusion in remote sensing carried out by the European Association of Remote Sensing Laboratories (EARSeL) and the French Society for Electricity and Electronics (SEE) [180]. According to this definition, data fusion is a *framework* containing means and tools for alliance or combination of data originating from different sensors. While fusion has been aimed at obtaining information of greater quality, the quality itself has been associated with the application. The definition of the word *fusion* has been compared with those of integration, merging, and combination. It has been suggested that these terms are more general, and have a much wider scope than fusion. It has also been argued that the term pixel does not have a correct interpretation as the pixel is merely a support of the information or measurement. The suggestions include usage of terms signal or measurement to describe the level of fusion. In this monograph, however, we use the term *pixel-level* to describe the same nomenclature as it has been followed by a large community.

Now, let us define a generalized model for basic pixel-based image fusion. Let I_1 and I_2 be two images of dimensions $(X \times Y)$ pixels having the same spatial resolution. Then the resultant fused image F is given by Eq. (2.1).

$$F(x, y) = w_1(x, y)I_1(x, y) + w_2(x, y)I_2(x, y) + C. \tag{2.1}$$

The quantities w_1 and w_2 indicate the relative importance assigned to the corresponding pixel at (x, y) of that image, and these are known as the fusion weights, or simply weights. As we want the fused image to be composed of the constituent images I_1 and I_2, we are essentially looking for an *additive* combination of these images. The fusion weights should also be ideally non-negative in nature. Additionally, the weights are normalized, i.e., the sum of all the weights at any given spatial location equals unity. Sometimes the constant term C is included, which acts as an offset or bias.

The most basic example of fusion is to average the input images. This leads to select the fusion weights as $w_1(x, y) = w_2(x, y) = 0.50 \ \forall (x, y)$ as per Eq. (2.1). The constant term is not used. It can be easily seen that this scheme is computationally most efficient. However, it fails to produce an output of the desired quality. The averaging technique explicitly assumes an equal amount of information to be present across the input images. In most examples, this is not the case. The IR image brings out very different information from the scene that does not get captured by a standard RGB camera. However, an average-based fusion would superimpose the features in the IR image by the RGB image, and thus, reducing the contrast and the information content. Therefore, an averaging-based fusion works well only when both the inputs are similar, and, lacks contrast when the inputs are different. This dissimilar information from multiple images causes a destructive interference which reduces the contrast. Therefore, despite its simplicity and computational efficiency, this method is rarely used in practice.

Fusion would be effective when the important spatial and radiometric features from the constituent images get retained, or appropriately enhanced during the process of fusion. Thus, one needs to extract the spatial features from images as the first step. In order to capture the unique features in input, Toet proposed the use of a Laplacian pyramid [172, 174]. The authors have proposed a hierarchical technique which decomposes each of the input image into a set of primitives defined by perceptually relevant patterns. This technique generates a pyramidal decomposition of each of the input images through filtering and subsampling the predecessor. The successive images in the pyramid are generally the reduced versions of the input image, and hence this representation is also referred to as the multi-resolution representation. The successive levels of the image pyramid represent image details and features with coarser approximations. The pyramidal representations of all the input images are then appropriately combined at every level using a pre-defined fusion rule. The fusion rule might be the same or different at every level, however, typically, one comes across two sets of rules. A fusion rule defined for all but the highest level in the pyramid is generally the same, and a different fusion rule is defined for the final or the topmost level image in the corresponding image pyramid. However it is possible to have a combination of more fusion rules. The combining process generates an image pyramid where each level represents the fusion of images at that particular level. The final resultant image can then be reconstructed by applying the reverse transformation on the fused image pyramid. Another popular pyramidal structure is obtained by convolving the current approximation of the image with the Gaussian filter. The pyramid so obtained is called a Gaussian pyramid. In [21], the filtering and sampling have been combined into a single operation resulting into the Gaussian weighted average. However, in [172], it has been argued that the linear filters alter the intensities of the pixels near the object boundary, and therefore, their applicability is limited when the precise measurements of the shape and size of the objects are needed. Their scheme employs a morphological multi-resolution decomposition of images using size-selective filters. It is claimed that morphological filters are more suitable for shape extraction of the objects in the fused image due to their property of removing the image details without adding any gray level bias.

Burt and Kolczynski [20] have proposed a generalized multisensor fusion using the gradient pyramid where the process of generation of the image pyramid has been referred to as the *pyramid transform*. Using the basis functions of gradient-of-Gaussian pattern, they apply the pyramid transform to both of the input images. Each of these basis functions is derived from the single prototype function via shifting, scaling, and rotation. This process decomposes the image up to three levels of successive approximations along four orientations. To combine the information across multiple decompositions, authors define two fusion rules. At the locations where the source images are similar, the fusion is achieved by averaging two images, while if the images are significantly different, the fusion rule selects the feature pattern with maximum saliency and copies it for the fused image.

Liu et al. have demonstrated the use of a steerable pyramid for fusion of remote sensing images [107]. The steerable pyramid is a multi-scale and multi-orientation decomposition with translation and rotation invariant sub-bands [167]. The low frequency or the coarsest approximation is fused based on the magnitude of the images at the corresponding locations. The fusion rule for the high frequency details is derived from the strength of each of the orientations which gives the directional information at that location. Authors refer to this selection rule as the absolute value maximum selection (AVMS).

Piella [135] has proposed a region-based technique in a multi-resolution framework as an extension of the pixel-based technique. This work provides a generalized structure for multi-resolution fusion techniques. The input images are first segmented which is a preparatory step toward the actual fusion. The author uses the term *activity measure* which captures the saliency in the image. The other quantity is the match measure which quantifies the similarity between the corresponding coefficients of the transformed images. This structure encompasses most of the pixel-based and region-based multi-resolution techniques, and also can be considered as the basis for the development of new ones.

For an efficient fusion, one needs to extract the salient features from multi-scale image decompositions. The wavelet transform has proved to be a highly popular tool for fusion. We assume that the readers are familiar with the fundamentals of the wavelet transform. For details on wavelets, one may refer to the dedicated texts by Daubechies [46], Vaidyanathan [178], Mallat [111], etc. The advantages offered by wavelets along with their theoretical background can also be found in [5, 124].

Li et al. have proposed a discrete wavelet transform (DWT)-based technique for fusion as wavelets offer distinct advantages such as orthogonality, compactness, and directional information [104]. Their technique is claimed to be superior to the Laplacian pyramid-based techniques as it does not produce any visible artifacts in the fused image as opposed to the later. Similar to the multi-resolution fusion approaches discussed earlier, the basic principle behind the wavelet-based fusion techniques is as follows—the set of input images is first decomposed into different multi-resolution coefficients that preserve image information. These coefficients are appropriately combined at each level to obtain new coefficients of the resultant image. This image is

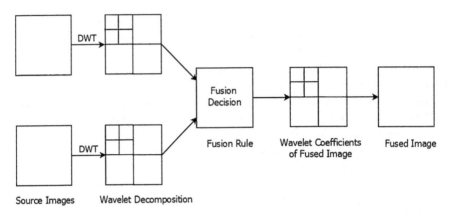

Fig. 2.2 Illustration of wavelet-based image fusion

then recovered via an inverse discrete wavelet transform (IDWT) to generate the final image. The schematic of the generalized DWT-based fusion techniques is shown in Fig. 2.2. The pyramidal representations of the input images are subjected to the fusion rule specified by the technique. This process generates the pyramid representing the decomposed fused image. The final image is obtained by an appropriate inverse transformation.

As stated earlier in this monograph, the key step lies in choosing an appropriate strategy to combine the coefficients, i.e., fusion rule. In [172, 174], the fusion rule has been defined to select the maximum of the corresponding coefficients of the ratio pyramid of input images, while the fusion rule that selects the maximum across discrete wavelet coefficients of the input images has been proposed in [104]. Mathematically, the general wavelet-based image fusion can be represented by Eq. (2.2).

$$F = \mathscr{W}^{-1}\left(\mathscr{F}\left(\mathscr{W}(I_1), \mathscr{W}(I_2), \ldots\right)\right), \tag{2.2}$$

where $\mathscr{F}$ is the fusion rule, and $\mathscr{W}, \mathscr{W}^{-1}$ are the forward and inverse wavelet transform operators, respectively. Wavelets have probably been the most successful family of fusion techniques. Wavelet-based fusion techniques have been implemented for various other application areas. Wen and Chen have demonstrated several applications of DWT-based fusion for forensic science [188]. Another application of wavelet decomposition for fusion of multi-focus images using the log-Gabor wavelets has been described by Redondo et al. [151]. The wavelet-based fusion techniques have also been proved to be useful for fusion of medical images. Performance of various multi-resolution techniques for fusion of retinal images has been analyzed in [96, 97]. In [86], wavelets have been shown to be useful for fusion of CT and MRI images.

Let us now take a brief look at some of the fusion techniques based on the variants of wavelets. A region-based fusion technique that generates a feature map through segmentation of the features of input images using a dual-tree complex wavelet transform (DT-CWT) to fuse images region by region has been discussed in [103]. The complex wavelet has also been effectively used for medical application related to fusion of CT and MRI images by Forster et al. [60]. De and Chanda have introduced morphological wavelets for fusion of multi-focus images [47]. The images are first decomposed using a nonlinear wavelet constructed with morphological operations. These decomposition operators involve morphological dilation combined with downsampling, while the reconstruction operators include morphological erosion combined with the corresponding upsampling. However, such a decomposition is not invertible. Later, these wavelet filters have been shown to be useful for CT-MRI fusion in [195], where the fusion rule is based on the selection of maximum absolute coefficients. Multiwavelets are the extension of wavelets which have two or more scaling and wavelet functions. Wang et al. have proposed a discrete multiwavelet transform (DMWT)-based fusion technique where they convolve the decomposed subbands with a feature extractor matrix to select salient features of the input images [181].

Multi-resolution-based fusion has been proved to be superior due to its ability in capturing information at different scales. During the last decade, several improvements have been proposed to capture features in an efficient manner. Scheunders has introduced the concept of multi-scale fundamental form (MFF) representation which provides a local measure of contrast for multivalued images [158]. On combination with a dyadic wavelet transform, it provides a measure of contrast in a multi-resolution framework for the description of edges. In [159], authors have defined a fusion rule that selects the maximum of the MFF coefficients of the input images. Chen et al. have improved this technique by weighting the MFF structure to avoid the enlargement of wavelet coefficients [35]. Mahmood and Scheunders have also applied an MFF-based strategy for fusion of hyperspectral images [109]. Petrović and Xydeas have proposed a representation of input images using the gradient maps at each level of decomposition pyramid as they improve the reliability of the feature selection [131, 132].

The 2-D separable transforms such as wavelets do not prove to be efficient in capturing the geometry of the images since the edges (or the points of discontinuity) may lie on a smooth curve, and thus, may not get accurately captured by the piece-wise linear approximation. Do and Vetterli developed the contourlet transform to capture the geometrical structure of images such as edges [50]. The contourlet transform involves the multi-resolution, local, and directional image decomposition. Fusion techniques based on multi-resolution contourlet transform have been proposed in [7, 114]. In [114], authors first obtain directional image pyramids up to certain levels (scales) using the contourlet decomposition. The low frequency coefficients which are at the top of the image pyramids are fused using the average-based rule. At the remaining levels, the fusion rule selects the coefficients from the source image which have higher energy in the local region. The fusion technique developed by Asmare et al. [7] also uses the averaging for combining the low frequency coefficients, while for the fusion of high frequency coefficients, it applies a match-and-activity based rule. The match and activity measures have been used to quantify the degree of similarity and saliency in the input images, respectively. The output of these measures is used to construct a decision map which in turn forms an actual weight of the fusion.

In [201], authors have explained the use of non-subsampled contourlet transform for fusion of multi-focus images. In this technique, low frequency coefficients have been combined with selection and averaging rules, while the high frequency coefficients have been selected on the basis of standard deviation at the corresponding level. In [197] the use of contourlet transform has been proposed for medical image fusion where the fusion rules include weighted average and local energy which are applied in the transform domain. The CT and MRI images have been fused using contourlet transform in [26] where the fusion weights are inversely proportional to the distance of the pixel from the current value of the fused pixel.

Curvelets have been another popular choice to represent edges [23, 52]. Curvelets are a multi-scale transform that more efficiently represent the edges and other singularities along curves than the wavelets. The efficiency refers to the ability to represent the data in a fewer number of coefficients for a given accuracy of reconstruction [52]. Choi et al. have experimented with curvelet transform for fusion of satellite images [40]. A finite ridgelet transform is obtained by taking a DWT of the coefficient vector of the finite Radon transform. Fusion of remote sensing images using the ridgelet transform has been proposed in [36].

A few researchers have applied the estimation theory to the problem of image fusion. The problem of fusion has been formulated as the following: The fused image is considered to be the underlying *true scene*. The input images obtained from multiple imaging sensors are assumed to depict the partial scene. The images contain an incomplete scene contaminated with some noise. The fusion problem thus gets transformed into the problem of estimation of the underlying true scene—which is the fused image itself. Once the relationship between the input images and the fused image is modeled, one can employ suitable techniques from the estimation theory which has a highly rich literature. Sharma [163], and Sharma et al. [164] have modeled the input images as noisy, locally affine functions of the true scene to be estimated. A Bayesian framework has been employed to obtain either the maximum likelihood (ML), or the maximum *a posteriori* (MAP) estimates of the true scene, i.e., the fused image. The parameters of this model have been estimated from the local covariance of the images. This model also offers flexibility to add a prior about the scene, if known; however, it assumes the noise component to follow a Gaussian distribution. Blum [15], and Yang and Blum [196] have improved the probabilistic fusion by allowing non-Gaussian noise distribution. An expectation maximization (EM)-based solution to detect concealed weapon through image fusion has been presented in [196]. For robustness against noise, a total variation (TV)-based prior has been incorporated into the probabilistic model by Kumar [94], and Kumar and Dass [95]. Xu et al. have proposed a Markov random field (MRF)-based prior for the fused image, and a Gaussian assumption of the parameters of the image formation model for hyperspectral image fusion [192, 193].

When the number of images to be fused is higher than three, the fusion problem can be viewed as a dimensionality reduction of the input data. Principal component analysis (PCA) is a powerful tool used for dimensionality reduction of multispectral images [150]. PCA transforms a set of intercorrelated variables into a set of new (potentially) uncorrelated linear combinations of new variables. The computation of principal components (PCs) of a 2-D image involves calculation of the eigenvectors of its covariance matrix. The PCA fusion procedure integrates the disparate natures of multisensor image data into one single image. If we assume the basis vectors, or equivalently the transform coefficients to be statistically independent, then we can apply the independent component analysis (ICA) to handle the fusion problem. In [116], the authors apply a DT-CWT technique of fusion in the bases constructed using ICA. Some methods of PCA-based hyperspectral image fusion are discussed in the next section.

Till now, we have discussed a number of fusion techniques mainly related to remote sensing applications. However, the spatial resolution of all input images undergoing fusion has been exactly the same. We explore in brief an important field of fusion that deals with images of different spatial resolution. In remote sensing, some of the satellites provide two types of images: a single band panchromatic (pan) image that represents the scene at a high spatial resolution, and a low resolution multispectral image containing a set of few bands. For example, the QuickBird imaging sensor provides a single band pan image of spatial resolution of 0.6 m, and a multispectral data containing 4-bands with a spatial resolution of 2.4 m. Similarly, the IKONOS captures panchromatic images of spatial resolution of 1 m, and a 4-band multispectral image with 4 m spatial resolution. The process of combining a set of low spatial resolution multispectral image with a co-georegistered high spatial resolution panchromatic image enhances the quality of multispectral image by increasing its spatial resolution. Fusion of pan and multispectral data produces a high-resolution pseudo-color image which preserves most of the attributes of both the image types, i.e., the finer spatial features of the pan image, and the spectral signatures from the multispectral bands. The fused image is, thus, spatially and spectrally enhanced, and hence visually appealing. This process which refers to the sharpening of a multispectral image using a panchromatic image is known as panchromatic sharpening, or simply pan-sharpening.

The intensity-hue-saturation (IHS) color space is considered to be quite useful for describing perception of colors to humans. The intensity represents the amount of brightness, the hue component refers to the color, and the saturation describes its purity. The appearance of the pan image is close to the intensity band of an IHS representation of the scene. Therefore, during fusion (pan-sharpening), the multispectral image is projected onto the IHS color space, and the intensity band is replaced by the pan image. The fusion output can then be obtained by the reverse IHS transform. A principal component analysis (PCA) transforms the intercorrelated data into a set of uncorrelated components. The first principal component (PC) can also be considered to be close to the pan data. The reverse transformation after replacing the first PC with pan data yields the sharpened multispectral image. The IHS- and PCA-based pan-sharpening techniques have been proposed in the early 1990s (cf. [33]). The contents of the pan and multispectral images are often dissimilar to a certain extent. Furthermore, these data differ in terms of radiometry (local mean) [61]. When these data are fused using the IHS-based methodology, it introduces a significant color distortion in the result [61, 202]. The PCA-based methods introduce less color distortion, but affect spectral responses of the multispectral data [61]. This spectral distortion is caused due to the mismatch of overlap between the spectral responses of the multispectral image, and the bandwidth of the pan image [61].

We have already discussed the utility of the multi-resolution based methodologies for fusion. This has also been used for pan-sharpening. The pan image is decomposed into a set of low-resolution images yielding a pyramidal structure. The low-resolution multispectral bands replace the pan image at an appropriate level of resolution from the pyramid. The reverse wavelet transform is then employed on each of the multispectral bands to produce the corresponding output. A non-orthogonal undecimated multi-resolution decomposition known as the "à trous" wavelet transform is most commonly employed by several pan-sharpening techniques [1, 4, 39, 149]. Aiazzi et al. have proposed a context driven thresholding of correlation coefficients between the images to be fused in the wavelet domain in order to avoid injection of undesired spatial details [1]. Zhang and Hong have integrated the IHS- and the wavelet-based techniques to reduce the color distortion [202]. In their technique, the multispectral bands are first projected onto the IHS color space prior to their wavelet decomposition. After the component substitution, the inverse wavelet transform followed by the inverse IHS transform generates the resultant sharpened image. In [66], the IHS transform of the multispectral image has been resampled to the size of the pan image. These images and the pan image are then decomposed using the Daubechies wavelet where the detail coefficients of the pan image are injected into the corresponding intensity component.

Contourlets have been known for better directional representation than wavelets, and capturing the geometrical structure of the objects [50]. Shah et al. have proposed the use of contourlets along with an adaptive PCA which preserves the spectral information [162]. Since the curvelets are more suitable for edge representation, they are also well suited for pan-sharpening. In [39, 40], curvelets have been used to extract the spatial details from the pan image.

Alparone et al. have proposed to extract the texture details from the pan image which are used to modulate the intensity of the multispectral image bands [4]. This technique is different from several others as it does not employ direct combination of image components. Joshi et al. compute a set of autoregressive (AR) parameters by modeling the spatial correlation of multispectral bands which are assumed to be the same for the pan image due to spectral correlation among them [83]. These parameters are then used in the process of regularization employed to combine the spectral characteristics of the multispectral image and the pan image. The generalized intensity component has been modeled as the weighted linear combination of the multispectral bands in [2]. These weights have been computed as the regression coefficients between the multispectral bands and the spatially degraded version of the pan image. The pan-sharpening has been carried out using the Gram-Schmidt spectral sharpening technique. Garzelli et al. have proposed a procedure to obtain a pan-sharpened image which minimizes the squared error between the multispectral data and the fused image [62]. Their technique provides the optimal results in the sense of the mean squared error (MSE). Moeller et al. have developed a wavelet-based pan-sharpening technique using hyperspectral images [118]. Their technique consists of a variational scheme that integrates objectives related to the image geometry, fidelity with respect to the pan image, and preservation of the spectral correlation.

2.2 Hyperspectral Image Fusion

We have discussed various techniques for pixel-based image fusion in the previous section, most of which have been proposed for fusion of a very few images (say 2–4). The extension of these techniques for fusion of hyperspectral data which contain 200–250 bands is a non-trivial task. While the fusion of hyperspectral images has been performed for various objectives such as quick visualization, resolution enhancement [72], spectral unmixing [208], etc., we restrict the scope of this section to the fusion techniques specifically developed for the scene visualization—which is the primary objective of this monograph.

The simplest way to fuse the hyperspectral data is to compute an average of all the bands. However, as mentioned earlier, this technique may provide satisfactory output when the input contents are highly correlated, but it results in the loss of contrast when the contents of the input are dissimilar. In the case of hyperspectral bands, the amount of independent information is quite high as one proceeds along the bands, and thus, mere averaging of the bands across spectral dimension lacks quality, although this technique is computationally most efficient. The distribution of information across hyperspectral bands is highly uneven [71]. Guo et al. have developed spectrally weighted kernels for fusion of hyperspectral bands for classification purposes [71]. The kernels which have been estimated using mutual information between image bands and the reference map of the same region, are in the form of Gaussian radial basis functions (RBFs). The fusion, however, is meant for classification purposes. Although, our primary objective is visualization of hyperspectral data, we believe that it is worth mentioning various attempts in the literature related to fusion of hyperspectral data.

Wilson et al. have extended a multi-resolution based technique for hyperspectral image fusion where contrast sensitivity defines the fusion rule [189]. Initially every band of the hyperspectral image is decomposed using a Gaussian pyramid. They have used a 5×5 Gaussian kernel formed through a convolution of two 3×3 Gaussian kernels to filter the input image as proposed in [20]. The size of the kernel indicates the spatial extent of the filtering operation. As the size increases, we are able to obtain a better filtering, but it increases the cost of computation. The filtered image is then downsampled by the factor of 2 at every stage to form a multi-resolution pyramid. Each level of this pyramid for each of the image bands is then subjected to the set of gradient filters to extract directional details (except for the topmost level which refers to the gross approximation). If $\mathcal{G}_k$ represents the pyramid at the decomposition level k, and ω is a 3×3 Gaussian kernel, then the directional decomposition for orientation l is given as $d_l * [\mathcal{G}_k + \omega * \mathcal{G}_k]$ $l = 1, 2, 3, 4$, in order to represent the gradient filters for detecting edges at $0°$, $45°$, $90°$, and $135°$, respectively [189]. The saliency is calculated by convolving the contrast sensitivity function with the magnitude of the low frequency Fourier components at each level for every orientation. The fusion rule is based on the perceptual contrast defined as the ratio of the difference to the sum of saliencies of input bands. The reconstruction of the fused image involves the following reverse procedure. The oriented gradient pyramid of the fused image is converted into an oriented Laplacian pyramid. This structure is then converted into a Laplacian pyramid through a sequence of filter-subtract-decimate (FSD) operations. The reduced Laplacian is then converted into a Gaussian pyramid from which the final fused image is constructed.

The problem of displaying the hyperspectral data onto a standard display device requires dimensionality reduction. Through fusion, one intends to retain the maximum features within the reduced data set. The principal component analysis (PCA) is a classic tool of dimensionality reduction which computes the basis vectors by analyzing the direction of maximum data variance, and projects the data onto them [150]. Tyo et al. have developed a PCA-based technique for hyperspectral image fusion where the mapping of the principal components (PCs) of the data is closely related to the human vision [177]. The first three eigenvectors of the covariance matrix of the input data represent the PC images in this case. To display the fusion result, the first PC has been mapped to the achromatic channel, while the second and the third PCs have been mapped to the R–G, and the B–Y color channels, respectively. The PCA-based method, however, turns out to be computationally very expensive when it is applied over the entire hyperspectral data cube. In [176], it has been suggested that the hyperspectral data should be partitioned before applying the principal component transform (PCT) emphasizing on the spectral properties of certain subsets of bands. Three strategies for partitioning are as following:

1. The basic strategy is to partition the data cube into three groups or subsets of equal size. This partitioning is based on grouping of successive bands, and referred to as the equal subgroup scheme by authors in [176]. The final RGB image is formed from the orthogonal band from each group.
2. The second strategy partitions the data to select the size of each subset using an iterative procedure such that the eigenvalues corresponding to the first PC of the subset become maximum [176]. At every iteration, the largest eigenvalue corresponding to the first principal component of each subgroup is calculated.
3. The aim of the partitioning is to reveal certain spectral characteristics of the data, so that the materials with significantly different spectral signatures should be distinguishable in the resultant RGB image. However, this scheme requires knowledge of the spectral signatures of the each material that is possibly present in the underlying scene.

Jacobson et al. have proposed the use of color matching functions derived from the hypothesis of how the synthesized image would be seen by the human eye if its range of perceived wavelength were to be stretched to the bandwidth of the hyperspectral image [79]. These functions, which act as the fusion weights specify the fractional components of the primary colors required to be mixed in order to create the sensation of the same color as that of viewing the original spectrum. For a K-band hyperspectral image, if $c_1(k)$, $c_2(k)$, and $c_3(k)$ represent the weights for the k-th spectral band related to the red, green and blue channels, respectively, then the fusion procedure to obtain the resultant RGB image $F(x, y) \equiv \{F_r(x, y), F_g(x, y), F_b(x, y)\}$ is given by Eq. (2.3).

$$F_r(x, y) = \sum_{k=1}^{K} c_1(k) I_k(x, y)$$

$$F_g(x, y) = \sum_{k=1}^{K} c_2(k) I_k(x, y)$$

$$F_b(x, y) = \sum_{k=1}^{K} c_3(k) I_k(x, y), \tag{2.3}$$

where $I_k(x, y)$ is the pixel at location (x, y) in the k-th hyperspectral band. These weights are subjected to the constraint—$\sum_{k=1}^{K} c_1(k) = \sum_{k=1}^{K} c_2(k) = \sum_{k=1}^{K} c_3(k)$ in order to satisfy the requirement of the equal energy white point [78]. The color matching functions have been derived from the CIE 1964 tristimulus functions, which are stretched across the wavelength axis so that the entire range of the hyperspectral image can be covered. For this stretching, only the first and the last bands of the hyperspectral image are equated with those of sRGB transformed envelopes, while the remaining values have been obtained via a linear interpolation [79]. The same authors have also introduced a set of fixed basis functions based on certain optimization criteria for display on the standard color space (sRGB), and the perceptual color space (CIELAB) [79]. These fixed basis functions are piecewise linear functions that provide a uniform variation in hue across the spectral response of the scene. The authors claim that the results produced using these functions are superior to the former case of the color matching functions in terms of providing a more uniform hue variation over the spectrum.

The problem of visualization of hyperspectral image has been formulated into an optimization framework in [44]. This work considers the goal of preservation of spectral distances. The process of fusion maps a hyperspectral image containing K bands into a color image of 3-bands. A pixel in the original data has been considered to be a K-dimensional quantity, while in the output, the same pixel has been represented as a 3-dimensional quantity. Here, the goal is to preserve the Euclidean norm of these distances from the input spectral space ($\in \mathbb{R}^K$) into the output spectral space ($\in \mathbb{R}^3$) in order to provide a perceptual interpretation of the scene. This method allows an interactive visualization where the user can modify the color representation of the resultant image. It uses an HSV color space which facilitates a faster remapping of the tone during interactive visualization. However, it is not as good as other color spaces such as L*a*b color space towards the preservation of spectral distances [44]. A spatial lens and a spectral lens have been provided as a part of interactive visualization, which allow users to enhance the contrast in a particular region, and to highlight pixels with a particular spectral response, respectively. Mignotte has also advocated the goal of preservation of spectral distances in [115]. This work has used the same measures as used in [44], albeit in the L∗a∗b color space L∗a∗b which is known to be perceptually uniform. It has been suggested that each of the objective functions to be minimized corresponding to one each in L, a, and b channels, can be defined as a Gibbs energy field related to a non-stationary Markov random field (MRF) defined on a complete graph formed with pairwise interactions of pixels. Having formulated the problem in a Markovian framework, it is possible to employ efficient optimization methods to obtain a minimum in the perceptual L∗a∗b color space. In [115], the images are decomposed into a 4-level multi-resolution pyramid, and the optimization is accomplished by coarse-to-fine conjugate gradient method.

In the previous section we have already explained the application of the matrix fundamental form (MFF) for image fusion [158, 159]. This scheme has been extended for the visualization of hyperspectral images in [109]. They perform denoising and contrast enhancement during the fusion of hyperspectral bands. This enhancement is achieved by combining the gradient information from the wavelet-decomposed hyperspectral bands. The low-resolution subbands, however, are combined using the color matching functions (CMFs) described in [78]. We have also explained in the previous section the Bayesian approach for hyperspectral image fusion proposed by Xu et al. [192, 193]. They have employed the statistical model of image formation where they have assumed the parameters to follow a Markov random field (MRF). They have also suggested modeling of the fused image using an MRF.

Cai et al. have proposed an interesting visualization technique for hyperspectral data with a different application—to display the results of mixed-pixel classification [22]. A single pixel in the hyperspectral data is a resultant of mixture of a number of materials present in the corresponding location on the earth. In the remote sensing literature, these materials are known as the endmembers, and their proportion into the pixel composition is known as the abundance [160]. The purpose of the methodology proposed in [22] is to provide a single image which displays—(*i*) overall material composition of the scene, and (*ii*) material composition at the particular location. The endmembers and their abundances are estimated using a linear mixture analysis which provides their spatial distributions. Each of the endmembers is then assigned to a different color, so that all of them can be visualized in a single color (RGB) image. The final color image has been composed of two layers. The first layer depicts the general distribution of the materials. The second layer provides the detailed composition for each pixel. The final image is composed by overlaying the detail-layer image onto the first one.

2.3 Quantitative Evaluation of Fusion Techniques

As the research in the field of image fusion began to develop, the problem of evaluation and assessment of fusion techniques has gained a lot of importance. Evaluation and assessment of fusion techniques are essential to compare different techniques, and to determine the benefits of each.

Quality of an image is an intangible subject for the common observer. Given two images, similar in content but significantly different in quality, one can easily recognize a relative quality. When the images are meant to be viewed by a human observer, the best method of quantifying the visual quality is through subjective evaluation [173, 184]. However, these quality measures depend on psychovisual factors, and therefore, such a subjective evaluation is difficult to reproduce and verify [136]. Furthermore, it turns out to be a too inconvenient, expensive, and time-consuming option. Therefore, it makes sense to develop a certain set of measures to assess the visual quality of an image objectively. The objective performance measures appear as a valuable complementary method to the subjective evaluation [136]. Wang et al. have enlisted three applications of designing an objective quality metric [184, 186]:

- to monitor image quality for control systems,
- to benchmark and compare image and video processing algorithms, and
- to optimize algorithms and parameter settings for image processing systems.

The image quality measures are broadly divided into two categories—(i) full-reference measures, and (ii) no-reference measures. In the former category, an original (noise- and distortion-free) image known as the reference image or the ground truth is available. The quality of an image (result of fusion in this context) is measured as its deviation from the original or the ideal image. An overview of such measures for assessing the fidelity of grayscale and multispectral images can be found in [9]. However, these measures require an ideal reference image. Practically it is not possible to have the *ideal* fused image. (Had it been available, fusion is no more needed!). Some performance measures quantify how closely the fused image is related to the constituent input images. Li et al. have composited an ideal-like image by manual cut-and-paste procedure from a set of multi-focus images [104]. However, this method is tedious, and is not applicable in general. Once the ideal, or the reference image has been defined, a metric such as root mean square error (RMSE) can be employed to compute the deviation, and thus, the quality as given by Eq. (2.4).

$$
\text{RMSE} = \left(\frac{1}{XY} \sum_{x} \sum_{y} (I_{\text{ref}}(x, y) - F(x, y))^2 \right)^{1/2},
\tag{2.4}
$$

where I_{ref} is the reference image. X and Y indicate the image dimensions. The measure of MSE (or RMSE) is also related to the peak signal to noise ratio (PSNR). These measures are easy to calculate, and are mathematically convenient to integrate into further processing. Although the MSE-based measures provide an elegant physical interpretation, these are not well correlated with the perceived quality of the image [185].

Xydeas and Petrović have proposed an objective measure to assess the quality of fusion based on the edge information [194]. This measure is quantified on the basis of *edge* information present in each pixel as it constitutes an important visual information for the human visual system. Their measure uses the Sobel edge operator

to yield the magnitude and orientation of edges at every pixel. The relative values of
the edge magnitude and orientation for every pixel are then calculated by comparing
those with the same for the corresponding pixel of the fused image. The relative edge
magnitude is calculated through the division of the edge magnitude of the input image
pixel and the fused image pixel, while the difference between the edge orientation
of the image pixel and the fused image pixel has been used to obtain the relative
edge orientation. These values are then used to derive the quantities referred to as
the edge strength and preservation. The actual expression for these values includes
additional constants that determine the exact shape of sigmoid functions which are
used to describe the edge strength preservation. The final measures for the magnitude
and the orientation model the loss of information in the fused image F, and quantify
how well the strength and orientation values of pixels in input images are being
represented in the fused image F. The product of these two final quantities is used
to evaluate the fused image, and thus, the corresponding fusion technique. When
this measure equals one, it indicates fusion with no loss of information from the
corresponding input image, while the value of zero refers to the complete loss of
edge information. However, this measure can quantify only the loss related to the
edge detection.

Wang and Bovik have proposed a universal image quality index ($UIQI$) for the
assessment of several image processing applications [183]. This measure is primarily
meant to evaluate the quality with respect to a standard or reference image. This
measure can be decomposed into three components—(i) correlation, (ii) luminance,
and (iii) contrast. The mathematical expression for $UIQI$ is given by combining the
second order statistical measures for the above three parameters. The $UIQI$ between
images I_1 and I_2 is calculated as:

$$UIQI = \frac{\sigma_{I_1 I_2}}{\sigma_{I_1} \sigma_{I_2}} \cdot \frac{2 m(I_1) m(I_2)}{||I_1||^2 + ||I_2||^2} \cdot \frac{2\sigma_{I_1}\sigma_{I_2}}{\sigma_{I_1}^2 + \sigma_{I_2}^2}, \tag{2.5}$$

where σ_{I_1} represents the standard deviation, $m(\cdot)$ represents the mean, and $\sigma_{I_1 I_2}$ is the
correlation between two images. The $\cdot$ operator is used to represent the dot product
between two lexicographically ordered images in the form of column vectors. It may
be seen that the final expression for the $UIQI$ can be simplified by canceling some
of the variance terms from Eq. (2.5). The dynamic range of $UIQI$ is $[-1, 1]$ where
the highest value of 1 indicates that the two images are exactly the same. Piella and
Heijmans have developed a metric for fusion evaluation based on the $UIQI$ [138].
They have proposed the use of a local window to calculate the $UIQI$ due to non-
stationary nature of an image signal. A weighted linear combination of these values
provides the required quality measure for the fused image. The weights are based on
some saliency measure in the local window.

When the pixel intensity can be treated as a discrete random variable, one can
effectively use some of the information theoretic measures to assess different image
processing operations. The entropy of a discrete random variable (image in our case)
refers to its information content. The mutual information (MI) represents the common

amount of information shared between two variables by measuring their statistical dependence. In the case of images, the MI describes the similarity in the distributions of intensities of corresponding images. Qu et al. have proposed the use of mutual information as the performance measure for image fusion [144]. If images I_1 and I_2 are fused to obtain the resultant image I_F, then the MI-based measure M_F^{12} is obtained using Eq. (2.6).

$$M_F^{12} = MI(F, I_1) + MI(F, I_2) \tag{2.6}$$

where $MI(F, I_1)$ and $MI(F, I_2)$ are the amount of mutual information between the fused image and corresponding input image. For the theoretical analysis of this measure the readers may refer to [37].

The calculation of these objective and statistical measures is easy when only a few images are to be fused. The expected behavior of these measures can be interpreted well—theoretically as well as intuitively. However, the extension of such measures for the assessment of hyperspectral image fusion is a non-trivial task. For example, although the definition of mutual information (MI) for a large number of variables is known, the computation and its physical interpretation is very difficult as the number of variables increases. Sometimes, the assessment is solely based on the quality of the output image itself. The correlation coefficient among the components of the final RGB image has been used to analyze the fusion technique in [49, 176, 206]. The spectral angle has been suggested as the distance measure between two vectors formed by two pixels in the hyperspectral data for evaluation of fusion in [78]. Wang et al. have proposed the correlation information entropy (CIE) to evaluate the performance of the fusion technique [182]. This measure quantifies the correlation between the images before and after fusion to determine the amount of information transferred from the source images to the fused image. Cui et al. have used preservation of spectral distance as a quality measure, which is evaluated over a sparse subset of image pixels to reduce computational requirements [44].

2.4 Notations Related to Hyperspectral Image

We have covered the current state-of-the-art in the area of hyperspectral image fusion. We have also familiarized our readers with several fusion methodologies for the general case of images. The next chapter onwards, we shall explain some recent techniques of hyperspectral image fusion in a detailed manner. Before we begin with the discussion, we shall first introduce to the readers the notations used in this monograph. These notations are consistent with most of the existing literature on hyperspectral images and image processing. In this monograph, we shall work with hyperspectral data which are 3 dimensional structures. These dimensions correspond to the spatial and the spectral information, respectively. Consider a hyperspectral image denoted by $\mathbf{I}$ of dimensions $(X \times Y \times K)$ where K indicates the number of bands with a spatial dimension $(X \times Y)$ each.

Fig. 2.3 Illustration of the notations for a hyperspectral image

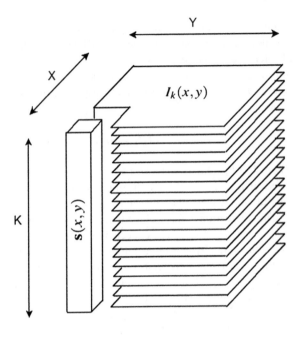

This image can be considered as a set of K two-dimensional spectral bands. We shall denote the k-th band by I_k, $k = 1, 2, \ldots, K$. Each I_k, therefore, refers to a 2-D spatial plane. The gray value of the pixel at the location (x, y) in this band is given by $I_k(x, y)$.

The hyperspectral image $\mathbf{I}$ contains spectral signatures of the pixels covered by the spatial extent of the imaging sensor. The image, therefore, is a collection of spectral signatures across locations (x, y), $x = 1, 2, \ldots, X$; $y = 1, 2, \ldots, Y$. Each of the spectral signature is a vector of dimensions $(K \times 1)$. A pixel in the fused image is primarily generated by processing the spectral signature of the pixel that represents an array of observations across all spectral bands for the given spatial location. We shall often work with this 1-D spectral array at a given location. Therefore, we shall introduce a different notation to refer to the spectral arrays (or vectors) in the hyperspectral image $\mathbf{I}$. We shall denote the array of observations (i.e., the spectral signature) at location (x, y) by $\mathbf{s}(x, y) \, \forall (x, y)$. It may be noted that this notational change is purely meant for the ease of understanding as the same data will be referred from different dimensions. Figure 2.3 illustrates the relationship between the two aforementioned notations. A hyperspectral image $\mathbf{I}$, thus, can be considered as a set of bands I_k where the set contains K elements. The same image can also be considered to be a set of spectral arrays $\mathbf{s}(x, y)$ where the set contains XY elements.

Chapter 3
Edge-Preserving Solution

3.1 Introduction

For efficient visualization of hyperspectral image, we would like to combine as many features from the spectral bands as possible. The features may be characterized by the edges, boundaries, or textures. The spectral response of some materials is possibly dominant only over a certain bandwidth. Various features in the data are thus clearly visible and observable only over a small subset of bands in the hyperspectral image. Typical pixel-level fusion techniques calculate the fusion weights based on some relative importance of the pixels, also known as the *saliency*. In this process, the features that are available across only a few bands may not receive adequate representation in the final image, if the weights are not chosen carefully. Thus, the weak features may get lost during the process of fusion. One would like to assign comparatively higher weightage to the pixels belonging to weak features in order to obtain a fused image where these features are appropriately represented.

The visual information is mainly available in the form of edges, lines, or boundaries—which constitute the high frequency components in the image. Most of the schemes of calculation of fusion weights or pixel saliency, therefore, employ some kind of high pass filtering (HPF) to extract the necessary information. The process of high pass filtering is equivalent to the subtraction of a suitably low pass filtered image from the original one. A conventional low pass filtering such as Gaussian filtering, however, distorts the edges in the image. If this low pass filtered image is used for the calculation of fusion weights, the edges and other sharp features do not get an appropriate representation, and hence the fused image contains visible artifacts. The edge-preserving filters are the class of non-linear filters which manipulate the image without distorting the boundaries and edges which are important elements of perceived visual quality. Edge-preserving filters have proved to be very useful in several applications of image processing and computer vision due to their ability to preserve edge information. This chapter explores an application of an edge-preserving filter known as bilateral filter for the fusion of hyperspectral image bands. We explain

S. Chaudhuri and K. Kotwal, *Hyperspectral Image Fusion*,
DOI: 10.1007/978-1-4614-7470-8_3, © Springer Science+Business Media New York 2013

how the edge-preserving property of a bilateral filter enables us to extract the minor features in the data, and obtain a visually sharp yet artifact free fused image.

Fusion of hyperspectral data often demands resources in terms of memory and time due to its huge volume. We demonstrate a hierarchical implementation of the bilateral filtering–based solution which overcomes the aforementioned issues. This scheme operates over smaller subsets of hyperspectral bands to provide the final resultant image without any degradation in the quality.

To begin with, we provide an overview of edge-preserving filters in Sect. 3.2, followed by a brief overview of bilateral filter in Sect. 3.3 which has been used to develop the edge-preserving fusion solution. The actual fusion solution has been explained in Sect. 3.4. The hierarchical scheme for an efficient implementation has been described in Sect. 3.5. In Sects. 3.6 and 3.7 we have provided the implementation details of the above solution, and the experimental results, respectively. Summary is discussed in Sect. 3.8.

3.2 Edge-Preserving Filters

Filtering is one of the most fundamental operations in image processing and computer vision. The term filtering refers to manipulating the value of the image intensity at a given location through a function which uses a set of values in its small neighborhood. When the images are assumed to be varying slowly over space, the adjacent pixels are likely to have similar values. As the noisy pixels are less correlated than the signal ones, the weighted averaging can minimize the effect of noise. However, the images are not smooth at the boundaries and edges. Therefore, the conventional smoothing or low pass filters produce undesirable effects at edges by blurring them.

The class of edge-preserving filters provides an answer to this problem by averaging within the smooth regions, and preventing this operation at the edges. Anisotropic diffusion implements the scale-space to generate a parameterized family of successively blurred images based on the diffusion equation [130]. Each of these images is a result of convolution of the image with a 2-D Gaussian filter where the width of the filter is governed by the diffusion parameter. This anisotropic diffusion process facilitates the detection of locations having strong edge information, and prevents them while smoothing. This technique is also referred to as the Perona-Malik diffusion. Black and Sapiro used the local statistics for calculation of a parameter related to the termination criteria of the Perona-Malik diffusion [14]. The use of a local noise estimate to control the diffusion process which is piecewise linear in nature for selective smoothing has been proposed in [105].

An adaptive smoothing operation can also be used to accomplish the similar goal. Adaptive smoothing involves repeated convolution of the image pixels with a very small averaging filter modulated by a measure of signal discontinuity at that location to produce smooth regions with preserved discontinuities [156]. Adaptive smoothing is a refinement of the Perona-Malik diffusion which is also related to the scale-space representation. An adaptive smoothing approach based on the higher order differentials, and the space-frequency analysis has been proposed in [25]. Chen combined the detection of contextual and local discontinuities for preservation of features during adaptive smoothing of image [34]. These quantities have been detected via the inhomogeneity and the local spatial gradient, respectively.

Bilateral filter is another edge-preserving non-linear filter. The kernel of bilateral filter is a product of the spatial domain kernel, and the range (or intensity) domain kernel [175]. The filtering operation, thus, weighs the closeness in intensity along with conventional spatial closeness. Therefore, the filtered images preserve edges which are characterized by dissimilar intensity contents. The SUSAN filter developed by Smith and Brady [169] is similar to bilateral filter which operates in the brightness and spatial domains for the purpose of edge detection, and thus, the structure-preserving noise reduction.

It may be noted that the aforementioned three approaches toward the edge preserving smoothing described in [130, 156, 175] are related to each other [11]. The relation between anisotropic diffusion and adaptive smoothing has been discussed [11]. It has also been shown that the extension of adaptive smoothing results in a bilateral filter. The next section describes the bilateral filter in detail.

3.3 Basics of Bilateral Filter

Bilateral filter was introduced by Tomasi and Manduchi [175]. This filter combines the pixel intensities for fusion based on their geometric as well as photometric closeness. The closeness of two pixels can be related to *(i)* nearby spatial location, and *(ii)* similar intensity/color values which are possibly useful in human perception. The basic idea behind bilateral filter is to employ an additional filtering kernel in range domain similar to a traditional filtering kernel in the spatial domain. The spatial domain filtering used in bilateral filter is essentially similar to the traditional filtering where the weights decay as one moves away from the pixel under consideration. In the range domain filtering, the weights decay with the dissimilarity in the intensities. That is, the weights are somewhat inversely proportional to the difference in the intensities. If the intensities of two neighboring pixels are very different, these pixels are likely to be parts of two different objects in the scene. These pixels, although may lie spatially close to each other, should not have much influence during the smoothing operation. This makes the filtering operation edge preserving, and eliminates only the finer textures while smoothing.

For an image $I(x, y)$ to be processed using a bilateral filter, let G_{σ_S} be the Gaussian spatial kernel similar to the traditional Gaussian filter. The spatial extent of the kernel is decided by the value of σ_S. Higher the spread parameter σ_S, the filter operates over a larger neighborhood of the pixel to be filtered, and more is the smoothing. Let G_{σ_R} be the Gaussian range kernel where σ_R decides how the difference in the intensities of the pixels gets converted into the fusion weight. If two pixels are parts of dissimilar regions, then these regions are probably separated by an edge. The range kernel parameter σ_R helps quantifying the strength of the edge in order to compute the range weight of the pixel. The value of σ_R thus controls the *amplitude* of the edge, and its corresponding weight.

If we denote the bilateral filtered image by $I^{BF}(x, y)$, then the filtering operation in discrete domain is given by Eq. (3.1).

$$I^{BF}(x, y) = \frac{1}{W(x, y)} \sum_{\tilde{x}} \sum_{\tilde{y}} \left\{ G_{\sigma_S}(x - \tilde{x}, y - \tilde{y}) \, G_{\sigma_R}\left(I(x, y) - I(\tilde{x}, \tilde{y})\right) I(\tilde{x}, \tilde{y}) \right\}$$

(3.1)

with,

$$G_{\sigma_S}(x, y) = \exp\left(-\frac{x^2 + y^2}{2\sigma_S^2}\right)$$

$$G_{\sigma_R}(\zeta) = \exp\left(-\frac{\zeta^2}{2\sigma_R^2}\right)$$

(3.2)

$$W(x, y) = \sum_{\tilde{x}} \sum_{\tilde{y}} \left\{ G_{\sigma_S}(x - \tilde{x}, y - \tilde{y}) \, G_{\sigma_R}\left(I(x, y) - I(\tilde{x}, \tilde{y})\right) \right\},$$ (3.3)

where $(\tilde{x}, \tilde{y})$ refers to the neighborhood of the pixel location (x, y). The term $W(x, y)$ is the normalization factor. For the 2-D spatial kernel G_{σ_S}, the spread value σ_S can be chosen based on the desired low pass characteristics. Similarly, for the 1-D range kernel G_{σ_R}, the spread value σ_R can be set as per the desired definition of what pixel difference constitutes an edge. In the case of an amplification or attenuation of image intensities, the σ_R needs to be appropriately scaled to maintain the results. However, due to the range filter component, the kernel of the bilateral filter needs to be recomputed for every pixel. The filter, thus, turns out to be computationally quite expensive.

The bilateral filter contains a range filter in addition to the conventional spatial filter kernel. The advantage of this additional kernel over the traditional spatial Gaussian filter can be illustrated through a simple example. Consider Fig. 3.1a that shows the 3-D representation of a step edge corrupted with an additive Gaussian noise. The bilateral filtered output of this noisy image (edge) smoothens the regions on the both sides of the edge, however produces a minimal degradation to the edge itself which can be observed from Fig. 3.1b. On the other hand, the conventional low pass filter smoothens the edge along with smoothening of other parts of the image as it can be seen from Fig. 3.1c.

The computational complexity is an important limitation of the conventional implementation of bilateral filter which is of the order of $\mathcal{O}(n^2)$ [175]. The bilateral filter involves a space-variant range kernel, which has to be re-computed for every pixel. However, in the past decade, several fast implementations of bilateral filter have been suggested. Paris and Durand have developed a signal processing-based implementation involving downsampling in the space and intensity which accelerates the computation by reducing the complexity to the order of $\mathcal{O}(n \log n)$ [126]. Another fast implementation using a box spatial kernel has been proposed by Weiss [187].

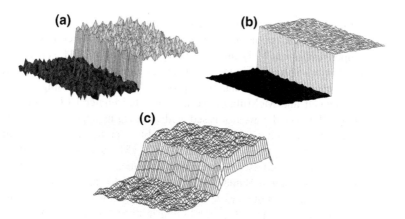

Fig. 3.1 Illustration of bilateral filtering for edge preservation operation. **a** A step edge and its noisy surroundings, **b** Output, and not output of the bilateral filter, and **c** Output, and not output of an averaging-based low pass filter

A real time bilateral filter with complexity $\mathcal{O}(1)$ has been designed by decomposing into a number of constant time spatial filters in [198]. Porikli has presented three methods for the computation of constant time bilateral filtering, i.e., with complexity $\mathcal{O}(1)$. These different implementations are based on polynomial filter expressions, linear filters, and a Taylor series approximation of bilateral filters, respectively [142]. Very recently Chaudhury et al. have proposed $\mathcal{O}(1)$ complexity bilateral filter using trigonometric range kernels [32].

In recent years, several improvements over the original bilateral filter in [175] have also been suggested. Pham et al. have proposed a separable implementation of bilateral filter which has been shown to be equivalent to the traditional filtering in terms of execution time [134]. An adaptive bilateral filter which tunes the parameters for every pixel based on its perceptual significance has been discussed in [190]. A trilateral filter which tracks the high gradient region in images by tilting the bilaterally smoothed image gradient vector has been proposed in [41]. Fattal et al. have applied a multi-scale strategy for improvement in the performance of bilateral filter [58]. The equivalence between bilateral filter, and anisotropic diffusion and adaptive smoothing has already been discussed in [11]. Barash and Comaniciu also proved that bilateral filter is an extension of the anisotropic diffusion process where the weights have been chosen with a certain geometrical considerations. Mean shift [43] algorithm can be employed in the joint spatial-range domains for discontinuity preserving smoothing. However, when the local mode search space is restricted to the range domain, it is equivalent to the bilateral filtering [12]. Elad has proved that the theoretical basis for bilateral filter lies in the Bayesian framework [55]. A recently developed non-local means filter [19] is a more generalized scheme where the small regions around the pixels are taken into consideration as opposed to the single pixel values in the case of bilateral filter during averaging.

Bilateral filter has been used in a variety of applications. To name a few, it has been used for classical problems such as deblurring from blurred/noisy image pairs [200], super-resolution [199], and optical flow estimation [101]. A 3-D mesh denoising application using this filter has been explained in [59]. A modified version- a separable bilateral filter has been proposed for pre-processing of video data in [134]. Bennett and McMillan have also adopted bilateral filter for video enhancement [13]. Jiang et al. have discussed medical imaging applications of bilateral filter [81]. Bilateral filter has received a considerable attention from an upcoming field of computational photography. Durand and Dorsey [54] have used bilateral filter to split images into detail and base layer to manipulate their contrast for the display of high dynamic range images. Raman and Chaudhuri [146] have proposed bilateral filtering–based extraction of weak textures from images which are in turn used to define the fusion weights. In [133], combining information from flash/no-flash photography using bilateral filter has been discussed.

3.4 Bilateral Filtering-Based Image Fusion

Let us consider a set of hyperspectral bands. We want to fuse these bands to generate a high contrast resultant image for visualization. We now explain the process of combining the bands with the help of edge preserving filter. Here we illustrate the fusion technique using a bilateral filter.

The primary aim of image fusion is to selectively merge the maximum possible features from the source images to form a single image. Hyperspectral image bands are the result of sampling a continuous spectrum at narrow wavelength intervals where the nominal bandwidth of a single band is 10 nm. (e.g., AVIRIS). The spectral response of the scene varies gradually over the spectrum, and thus, the successive bands in the hyperspectral image have a significant correlation. Therefore, for an efficient fusion, we should be able to extract the specific information contained in a particular band. When compared with the image compositing process, our task is analogous to obtain the mattes, popularly known as α-mattes, for α each of the source images. The mattes define the regions in the image, and the proportion in which they should be mixed. These mattes act as the fusion weights to generate the final image having the desired features from the input. The final image F can be represented as a linear combination of input images I_k, $k = 1$ to K as shown in Eq. (3.4).

$$F(x, y) = \sum_{k=1}^{K} \alpha_k(x, y) \, I_k(x, y) \quad \forall \, (x, y), \tag{3.4}$$

where $\alpha_k(x, y)$ is the α-matte for the pixel at location (x, y) in the k-th observation. The techniques of automatic compositing employ a matte generating function as the function of input data itself. We have developed the fusion strategy by choosing the weights in the spirit of such α-mattes. This methodology is similar to the compositing of high dynamic range-like photographs with variable exposures as discussed in [146]. Our focus, however, is to handle the problem of fusing remote sensing images with variable reflectance of different regions contained in the scene spread over a wide spectrum.

Like any other pixel-based fusion technique, the output image is generated as a normalized weighted sum of the pixels from the input bands at the corresponding location. The critical part of the algorithm is computing appropriate weights to represent the subtle information at each location along the spatial and the spectral dimension. We define the weight of the particular pixel from its relative importance with respect to its spatial neighborhood. Hyperspectral images are generated as the reflectance response of the scene, which mainly depends on the composition of the materials in the scene. Certain materials exhibit a stronger response over a given wavelength range, and give rise to strong and sharp features. On the other hand, some materials exhibit their peak response over a very narrow spectral range. Thus, the data contain several weak features such as edges, which are prominent over only a set of few bands, while the strong features appear in a large number of bands in the data. When the final image is a linear weighted combination of the input images, the strong features get preserved due to their significant, non-zero weights almost in every band. The weak features, however, may get lost due to their local presence across the spectral dimension. We want to specifically focus on preservation of these weak features. Therefore, we assign higher weightage to those regions in the band where the weak features are strong so that these will be well perceived in the output.

The first step in this regard is to identify the features in the form of weak edges, fine textures, and objects apparent over only a few bands. Such slowly varying features of hyperspectral band can be removed by a smoothing 2-D filter, which when subtracted from the original image give the important local details of the band. However, a conventional Gaussian low pass filter tends to blur the edges, and the difference image so formed contains artifacts in the neighborhood of the edge pixels. This leads to visible artifacts in the fused image. Therefore, we need a smoothing filter that removes minor variations in the image, but does not smooth out the strong edges. Bilateral filter, discussed in the previous section, satisfies our requirements for the design of mattes, which in turn generate the fused image.

Having explained why the weak features should be given an adequate importance while defining the matting/weighing function, let us now illustrate how bilateral filter serves this purpose of extracting weak features through Fig. 3.2. A small region of an urban area in Palo Alto has been depicted in Fig. 3.2a. This is a part of the 50-th band of the hyperspectral image captured by the Hyperion imaging sensor. Fig. 3.2b shows the output of bilateral filtering over the original data. As it can be easily observed, this has retained the strong edges, however it has removed weak features, and it appears slightly blurred. The difference image (between the original and the filtered) in Fig. 3.2c, thus displays these weaker parts in the original band more prominently.

The difference image in Fig. 3.2c depicts the detail layer or the component of the image which has been removed by the filtering operation. We formulate our weights as the function of the difference image. Our weights are, thus, directly proportional to the finer details in the given hyperspectral band. Let $I_1, I_2, \ldots I_K$ be the set of

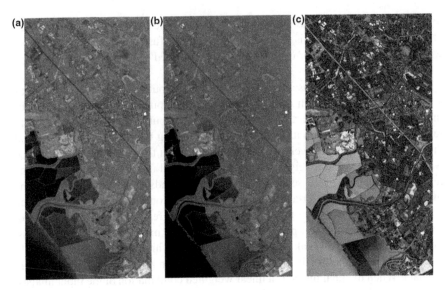

Fig. 3.2 Illustration of bilateral filtering for the purpose of detail extraction. **a** Depicts an original band of hyperspectral image, **b** shows the output of the bilateral filter, and **c** shows the absolute difference image between **a** and **b**. The image **c** has been contrast enhanced for the display purpose

hyperspectral image bands, containing K consecutive bands. We calculate fusion weights $w_k(x, y)$ as the following:

$$w_k(x, y) = \frac{|I_k(x, y) - I_k^{\mathrm{BF}}(x, y)| + C}{\sum_{k=1}^{K}(|I_k(x, y) - I_k^{\mathrm{BF}}(x, y)| + C)}, \tag{3.5}$$

where I^{BF} is the corresponding filtered band. C is a positive real number which serves three purposes:

- It provides sufficient weightage to the strong edges which are absent in difference images. The strong edges are least affected by the bilateral filter, and hence their presence is minimal in the filtered image which acts as the weights. The constant C provides a constant weightage which acts as the bias along with the actual fusion weight computed from the bilaterally filtered image.
- It brings flexibility to the process by controlling the effect of actual weights w. When the value of the constant C is very small, its effect on fusion can be practically neglected. In such case, the actual weights dominate the fusion process, and the resultant image, thus, purely reflects the strength of the locally dominant features. On the other hand, when the constant is a significantly large number as compared to the fusion weights, the fusion becomes an averaging process. Through selection of an appropriate constant C, the user can adjust the nature of the fused image.

- The homogeneous regions in the image remain unaltered after filtering. The fusion weight in such areas is effectively zero. If any image region appears to be homogeneous over all the bands in a hyperspectral image, the denominator in the Equation of fusion weights (Eq. (3.5)) turns out to be zero in the absence of a constant C. This causes numerical instability in the formulation of fusion process. Incorporation of the numerical constant ensures that the denominator is always a non-zero, positive number, and thus, the fusion weights are generated within a proper range. The constant, thus, prevents numerical instability in the homogeneous regions.

The denominator of Eq. (3.5) guarantees that the sum of all weights at any spatial location equals unity, i.e., normalized weights.

$$\sum_{k=1}^{K} w_i(x, y) = 1 \quad \forall(x, y).$$

The final fused image of the hyperspectral data is given by Eq. (3.6) as,

$$F(x, y) = \sum_{k=1}^{K} w_k(x, y) I_k(x, y). \tag{3.6}$$

3.5 Hierarchical Implementation

We are dealing with the problem of combining a few hundred bands into a single image, either a grayscale or an RGB. The process may involve reading all the bands in the input hyperspectral image at once, computing the weights, and generating a resultant fused image as the linear combination of all the input bands. This *one time* reading and combining all the image bands have the following shortcomings.

1. This results in assigning very small fractional weights to the locations in each of the image bands. Some of the weights are even comparable to the truncation errors which might lead to washing out some of the minor details.
2. It requires the entire data along the spectral dimension to be read. Therefore, the entire hyperspectral data cube has to be read into memory. Due to the huge size of a hyperspectral data, the memory requirement is over a few hundreds of megabytes.

We describe a hierarchical scheme for fusion in order to overcome these issues. Instead of employing the bilateral filter-based scheme to the entire set of bands, this scheme creates partitions of the data into subsets of hyperspectral bands. For a given image of dimensions $(X \times Y \times K)$, containing K bands, one forms P subsets of dimensions $(X \times Y \times M)$ from contiguous bands of the data, where P is given by $P = \lceil \frac{K}{M} \rceil$. Then, the bilateral filter-based fusion scheme may be employed to

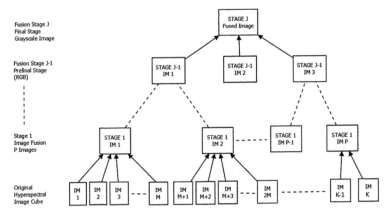

Fig. 3.3 Illustration of the hierarchical scheme for hyperspectral image fusion. (©2010 IEEE, Ref: [88])

each of these subsets independently to obtain P fused images. Let us call them first-stage fusion results. These P images form the base images (or inputs) for the second-stage fusion. Now we may apply the same fusion algorithm over this image cube of dimensions $(X \times Y \times P)$. This procedure is repeated in a hierarchical manner to generate the final result of fusion in a few stages. If we generate three such images during the pre-final stage, and assign them to different color channels, we obtain the RGB representation of the hyperspectral image.

Figure 3.3 shows the schematic of the hierarchical scheme of fusion. The scheme begins with the formation of P subsets of hyperspectral bands each containing M bands each, except the last one which may contain slightly less number of bands depending upon the total number of bands K in the hyperspectral image. The first-stage of fusion generates P intermediate fused images, one from each of the subsets of the original data. The same process continues till the single fused image which represents the combining of the entire data is generated. If the complete process operates at J levels, at the pre-final stage, i.e., the $(J - 1)$-th stage, we obtain three independent images each of them representing fusion of approximately one-third dataset. We can create an RGB version of the fused image by assigning these three images to red, green, and blue channels, respectively. The scheme brings flexibility in terms of number of bands undergoing fusion. In literature, multi-resolution based fusion techniques are sometimes referred to as the hierarchical fusion. The term hierarchical has been used in a different context in this chapter. This scheme does not perform any multi-resolution decomposition of the image/bands.

Typically, if we select M to be nearly 10 % of the original number of bands K, then only those many bands are required to be present in the memory at any time. The value of M, however, is subject to the choice of the observer, and does not much affect the quality of the final result when the number of images being fused at any stage are fairly equal. Although we have provided the illustrations by choosing the bands in forward succession, the bands can also be combined in any random order for the given hierarchical implementation. However, the successive bands in the hyperspectral data exhibit a high degree of similarity. When the randomly arranged bands are grouped for the first level of hierarchical implementation, the corresponding fusion results contain integration of highly dissimilar information as compared to the sequential case.

The advantages of the hierarchical scheme are as follows:

- This scheme reads only a fraction of the hyperspectral image bands at a given time to generate an intermediate fusion result. The hierarchical scheme requires only a small subset of the input to be read in the memory and process it. Thus, the memory requirement is significantly reduced.
- It enables the user to scale up the system to efficiently fuse any number of images. It can easily accommodate any increase in the number of bands without compromising with the performance.
- The system is computationally efficient, and makes it possible to parallelize the implementation. All the subsets of the hyperspectral image can be processed in parallel to produce a set of intermediate fusion results.
- The resultant images at the intermediate stages facilitate analysis and visualization of midband reflectance response of the scene. The fused images at the first-stage represent the response of the scene over a bandwidth that is M times that of an individual hyperspectral band. These and the subsequent intermediate results can be used to visualize the combined response of the scene over a range of bandwidth encompassed by the number of bands being fused.

3.6 Implementation

The fusion procedure requires selection of three parameters- σ_S, σ_R, and C. The choice of these parameters is important to achieve a better quality of the output. The implementation of the bilateral filter is also an important parameter as far as the computational complexity and timing requirements of the entire procedure are concerned. Here we use the implementation based on the approximation of bilateral filter provided in [126]. In order to automate the fusion algorithm without much degradation in quality, we have adopted the guidelines suggested in [126] to select the values of first two parameters.

$$\sigma_S = C_1 \times \min(X, Y) \tag{3.7}$$

$$\sigma_{R_k} = C_2 \times \left(\max(I_k(x, y)) - \min(I_k(x, y)) \right) \quad \forall k, \tag{3.8}$$

where C_1 and C_2 are positive real numbers which can be set to obtain the desired quality output. The choice of C_1 is related to the size of spatial details retained during fusion. We have used $C_1 = 1/16$ in all test experiments. The value of the range kernel defines the minimum amplitude that can be considered as an *edge*. We have set C_2 to 0.05 in our experiments.

A smaller value of the constant C in Eq. (3.5) results in boosting the finer details which may result in over-sharpening. On the contrary, a too high value of C smoothens out fusion results which makes it *average-like*. We have selected the value of C to be equal to 50 in this experiment while the hyperspectral bands are a 16-bit data.

In the case of the RGB version of the fusion result, the three pre-final stage images have been assigned to the red, green, and blue channels of the displaying device. The RGB visualization of hyperspectral images inherently involves pseudo-coloring, and hence, the coloring scheme does not bear any physical meaning.

3.7 Experimental Results

The experimental results of the presented techniques are typically provided in the same chapter for most of the books. However, the subsequent chapters of this monograph present three more techniques for visualization of hyperspectral images. We would like to provide a detailed comparison of these techniques together which, we believe, would facilitate a quick and easy analysis of these fusion solutions to the reader. Therefore, instead of providing separate results for each of the techniques, we have provided the results of all the techniques presented in this monograph, as well as some other existing hyperspectral image fusion techniques together in a specially dedicated Chap. 10 at the end. Also, these results will also involve a quantitative performance evaluation of these techniques using various objective performance measures. We would like the readers to be familiar with these performance measures before we discuss merits and demerits of the fusion techniques on the basis of these measures. In Chap. 9, we explain in detail various performance measures that quantify various aspects related to fusion of a large number of images. In the present chapter, however, we shall illustrate in brief the experimental results for fusion over just a couple of datasets.

We demonstrate the efficacy of the presented solution using a real world hyperspectral data provided by the Hyperion imaging sensor used in the EO-1 spacecraft for the earth observation. The dataset consists of 242 bands (0.4–2.5 μm) with a 30 m spatial resolution. The selected data set depicts the urban region of Palo Alto, CA. The dimension of the Hyperion data cube is $(512 \times 256 \times 242)$. This dataset is a part of third sub-block of the original dataset. We refer to this dataset as the urban dataset throughout this monograph. We have employed a three-stage hierarchical scheme for fusion over this dataset. The fused images in the first stage comprise a bandwidth of approximately 110–120 nm, which are generated by fusing 12 contiguous bands. The second-stage fusion generated three images from 18 resultant images of the first-stage fusion. The first fused image as shown in Fig. 3.4a results from fusing bands 1–74, corresponding to the wavelengths from 350.505 nm to 887.235 nm. The second fused image [Fig. 3.4b] of the pre-final stage represents the fusion over bands 75–154. The bandwidth of this fused image of the urban data scene is 807.531 nm (from 887.240 to 1694.771 nm). The last fused image at this stage has a bandwidth of 888.508 nm, which represents the fusion of bands 155–242 of the original hyperspectral image

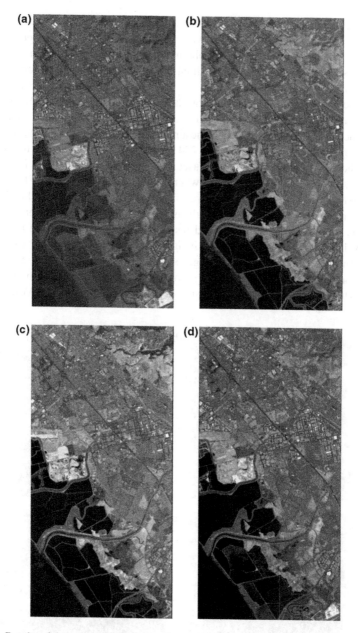

Fig. 3.4 Results of the second-stage fusion of the urban image cube from the Hyperion, and the final fused image. **a** Fusion over bands 1–74. **b** Fusion over bands 75–154. **c** Fusion over bands 155–242. **d** Final fused image. These images are linearly scaled to the range [0, 255] with gamma set to 1.50 for display purposes. (©2010 IEEE, Ref: [88])

cube, corresponding to the wavelengths from 1693.765 to 2582.273 nm. Fig. 3.4a–c show these fused images of the Palo Alto region collected by the Hyperion imaging sensor after the second stage. We obtain the final grayscale fused image of this urban dataset by fusing these three pre-final stage images, which has a bandwidth of 2231.768 nm and is shown in Fig. 3.4d.

Let us consider another hyperspectral dataset provided by the AVIRIS imaging sensor developed by the Jet Propulsion Laboratory (JPL/NASA). This is an airborne sensor where the data get collected as the aircraft flies over the corresponding region of the earth. This dataset consisting of 224 bands depicts the region of the Moffett Field in the CA. Let us refer to this dataset as the moffett$_2$ dataset. This dataset has the dimension of $(614 \times 512 \times 224)$ where each pixel is represented by 16 bits. Similar to the previous example, we have employed a three-stage hierarchical fusion scheme for the fusion of the moffett$_2$ dataset. Each of the resultant images of the first stage, thus, represents a combined response of the scene over the bandwidth of nearly 120 nm where each of these images are contiguous over the wavelength spectra. These 18 first-stage fused images are evenly grouped, and fused using the same fusion methodology to form three second-stage fused images. The first of these images, which is actually a result of fusion of the first 78 bands of the AVIRIS data

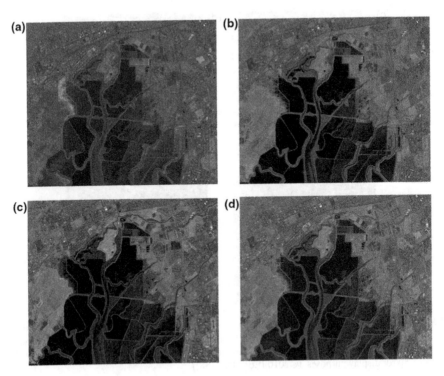

Fig. 3.5 Results of the second-stage fusion of the moffett$_2$ image cube from the AVIRIS, and the final fused image. **a** Fusion over bands 1–78. **b** Fusion over bands 79–152. **c** Fusion over bands 153–224. **d** Final fused image. (©2010 IEEE, Ref: [88])

comprising a bandwidth of 735.190 mm, has been shown in Fig. 3.5a. The subsequent resultant image as shown in Fig. 3.5b represents fusion of the next one-third block of the moffett$_2$ dataset corresponding to the bands from 79–152 of the original data. The total bandwidth encompassed by these bands, and thus, by the corresponding fused image in Fig. 3.5b is 706.875 nm. The remaining subset of the hyperspectral bands forms the input for the third fused image at this stage as shown in Fig. 3.5c. It may be noted that these bands have been initially fused into six first-level fused images, which are then combined to obtain the second-level fused image. This image essentially combines the remaining 72 bands with a total bandwidth of 705.140 nm. Fig. 3.5d depicts the final fused image obtained through fusion of three resultant images from the pre-final stage, i.e., the second-level fusion stage in this case. This image has the total bandwidth of 2146.780 nm which is the sum of the individual bandwidths of the constituent images at any given stage.

3.8 Summary

This chapter explains a solution to fusion of hyperspectral image for the purpose of visualization. This solution uses an edge preserving bilateral filter for defining the fusion weight based on the locally dominant features. The discussed solution provides flexibility in the process of fusion by modifying the bilateral filter parameters, and the tuning parameter C to match the needs of the observer. As the bilateral filter is non-iterative and efficient implementation schemes are available, this solution is fast and computationally simple. Furthermore, this technique does not require construction of multi-resolution pyramids. As it requires the fusion weights to be computed for each and every pixel of the hyperspectral data cube, the fusion process exploits all available information at all locations in all bands, unlike the techniques where the entire band receives the same weight.

This chapter also discusses a hierarchical implementation scheme for fusion so as to accommodate any increase in the number of bands without degradation in fusion performance. Additionally, this scheme enables visualization and analysis of fusion of bands up to any given spectral bandwidth.

Chapter 4
Band Selection Through Redundancy Elimination

4.1 Introduction

Pixel-based image fusion techniques compute the fusion weights for each pixel in every band. Thus, every band in the hyperspectral data is subjected to the process of computing weights through some kind of saliency measure which is a time-consuming operation. Let us divide the process of fusion into two steps- first, computation of fusion weights through saliency measurement, and second, a linear combination of the bands using these weights. The later step is quite trivial, and requires a very little computation. The first step, however, is critical to the performance of the algorithm, and can be computationally demanding as per the chosen algorithm. While some attempts towards a quick visualization of the image contents in the form of an RGB image have been investigated, some of these methods typically include selection of three image bands satisfying certain criterion [49, 79]. However, these approaches select only three bands for the display, and they do not involve any kind of image merging or fusion at any level.

For the data (or pixel) level image fusion, most available techniques evaluate the importance of a pixel within its spatial neighborhood, and then assign appropriate weights to the pixels while fusing them over various image bands. For example, in the previous chapter, the residual image after subtraction of the image band from its bilateral filtered version provides the necessary fusion weights. Since this step consumes the major amount of computation, usually on a per pixel basis, the time taken for fusion is directly proportional to the number of image bands. An observer has to wait until the completion of calculation of weights for the entire set of image bands, followed by the successive weighted addition to get the final result of fusion. Therefore, the fusion techniques tend to be slower due to the large number of image bands. The process of calculation of weights can be computationally quite demanding and time-consuming when the fusion techniques and /or the saliency detection techniques are iterative in nature. This often limits us from exploring some sophisticated but computationally demanding methods (such as the one given in Chap. 6) for hyperspectral image fusion.

S. Chaudhuri and K. Kotwal, *Hyperspectral Image Fusion*,
DOI: 10.1007/978-1-4614-7470-8_4, © Springer Science+Business Media New York 2013

In this chapter, we shall discuss a method for selection of a few specific bands to accomplish a faster fusion of hyperspectral images, without much sacrificing the quality of the result of fusion. We present an information theoretic strategy for choosing specific image bands of the hyperspectral data cube using only the input hyperspectral image. The selected subset of bands can then be fused using any existing pixel-based fusion technique as the band selection process is independent of the method of fusion. The objective of this chapter lies in providing a much faster scheme for hyperspectral image visualization with a minimal degradation in the fusion quality through selection of specific bands. The band selection process should be computationally inexpensive, and yet, should be able to generate comparable quality fusion results for the given fusion technique.

The next section describes a possible scheme of entropy-based band selection. This is a generic scheme for the data containing a large number of bands. The hyperspectral data consist of a spectrally ordered set of contiguous bands. We exploit this *order*-based characteristic, and develop a model for measurement of the similarity across bands. Section 4.3 provides details of the model, and develops the special case for spectrally ordered hyperspectral data—which is usually the case. We also provide a couple of theorems for the savings in computation for this special case of band selection. Section 4.4 consists of the experimental results and the performance analysis of the band selection scheme. Summary is presented in Sect. 4.5.

4.2 Entropy-Based Band Selection

In case of hyperspectral imaging sensors, these images are acquired over a narrow but contiguous spectral bands. Therefore, several image bands exhibit a very high degree of spectral correlation among them, typically highest degree of correlation is found within two successive bands as they depict the reflectance response of these scene over contiguous, adjacent wavelength bands. For instance, consider the urban hyperspectral image acquired by the Hyperion, and the moffett$_2$ hyperspectral image acquired by the AVIRIS sensors. We may measure the correlation among the adjacent bands of the respective datasets using the correlation coefficient CC, which is defined by Eq. (4.1).

$$CC(k) = \frac{\sum_{x=1}^{X} \sum_{y=1}^{Y} (I_k(x, y) - m(I_k)) (I_{k+1}(x, y) - m(I_{k+1}))}{\sqrt{\sum_x \sum_y (I_k(x, y) - m(I_k))^2 \sum_x \sum_y (I_{k+1}(x, y) - m(I_{k+1}))^2}} \quad 1 \leq k < K,$$

(4.1)

where $I_k, k = 1$ to K represents the bands of the hyperspectral data. The $m(.)$ operator indicates mean of the corresponding argument. Figure 4.1 shows the correlation coefficient among the adjacent bands of the urban and the moffett$_2$ datasets. Noisy bands have been already discarded. The average value of the correlation coefficient for both the datasets is higher than 0.97. When a fusion algorithm operates over the similar bands, a very little amount of additional information is contributed towards the fusion result. Therefore, one may think of selecting a subset of a fewer number

Fig. 4.1 Correlation coefficient (*CC*) over the adjacent bands in the hyperspectral data. The *black line* represents the *CC* for the urban data from the Hyperion, and the *blue line* represents the *CC* for the moffett$_2$ data from the AVIRIS

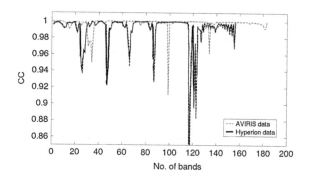

of hyperspectral bands that contain a higher amount of independent information, and fusing this subset of bands instead of processing the entire set of hyperspectral bands. Fusion of such a subset would generate an output image without much loss in the quality when compared with the output image obtained from fusion of the entire data using the same fusion technique. We explore this possibility of selection of subset of bands where each band possesses a certain amount of additional information with respect to other bands, and thus, together they capture of the information content in the hyperspectral image cube. We describe a conditional entropy-based scheme for the selection of a fewer number of image bands which are mutually less correlated in order to facilitate a faster visualization. This scheme turns out to be quite faster and memory efficient as one can achieve the required quality of fusion using only a much smaller subset of the hyperspectral data.

4.2.1 Redundancy Elimination

An image band undergoing fusion should possess a significant amount of additional information for the fusion process to be efficient. We shall now explain an algorithm to select a subset of image bands based on conditional entropy. Let $\mathbf{I} = \{I_k; \ k = 1, 2, \ldots, K\}$ be the hyperspectral image consisting of K bands. We seek to identify only a subset of bands that will actually undergo fusion. We denote this subset of cardinality $\bar{K}$ as $\bar{\mathbf{I}}$, such that $\bar{\mathbf{I}} \subset \mathbf{I}$, and $\bar{K} \ll K$. The first band is trivially selected for fusion, which forms the first element of the subset of bands to be fused, i.e., $\bar{\mathbf{I}} = \{I_1\}$. The conditional entropies of the successive bands with respect to this band are evaluated. The next band is selected when the corresponding conditional entropy exceeds a pre-determined threshold, i.e., when the additional information content in the given band is sufficiently high. This threshold has been selected as an appropriate fraction of the entropy of the band under consideration. The newly selected band becomes a member of $\bar{\mathbf{I}}$ subset. Subsequently, for every image band, the conditional entropy with respect to each of the previously selected bands $\bar{I}_k$ is calculated. The motivation lies in the fact that a newly available image band may not be well correlated with the last of the selected subset of bands, but it may be well correlated with one of the bands selected earlier. In either case, the newly available band should be considered as redundant for fusion purposes. The band is selected for fusion if it has a conditional entropy higher than the threshold. Thus, given a set of hyperspectral images $\{I_k;\ k = 1, 2, \ldots, K\}$, the p-th image band selected for fusion is given by Eq. (4.2).

$$p = \arg \min_r \left\{ \min_k H(I_r | \bar{I}_k) \geq \Theta,\ r > k \right\}, \qquad (4.2)$$

where $\bar{I}_k$ represents the subset of selected bands up to the k-th available band in $\mathbf{I}$, $H(I_r | \bar{I}_k)$ represents the entropy of the image I_r conditioned on each of the element of the set $\bar{\mathbf{I}}$.

The threshold Θ is empirically chosen as a suitable fraction of $H(I_r)$, i.e.,

$$\Theta = \kappa\, H(I_r), \qquad 0 < \kappa < 1. \qquad (4.3)$$

This procedure is continued until the entire dataset is exhausted. Any pixel based fusion scheme can then operate over this selected subset $\bar{\mathbf{I}}$ to generate an appropriately fused image.

The aforementioned scheme exploits the statistical redundancy in the input data. It discards bands that are very similar to the selected ones, and selects the ones that are quite different in terms of additional information content. Therefore, although a fewer images are selected, most of the information content in the data is captured by the band selection process. The resultant fused image, thus, contains most of the features of the entire dataset.

This scheme essentially selects the image bands by evaluating the additional information content against all the previously selected bands. This scheme does not take into consideration any possible pattern of the correlation among the bands. Therefore, the band selection scheme is guaranteed to yield good results for any organization of input data consisting of multiple images of the same scene to be fused. However, a band selected once cannot be removed from the subset $\bar{\mathbf{I}}$ throughout the process; thus this scheme is based on the greedy selection of image bands.

4.3 Special Case: Ordered Data

Consider a typical problem of hyperspectral image data to be fused for efficient visualization. The hyperspectral data consist of a set of narrow-width, but spectrally contiguous bands acquired by an array of sensors. As the bands are ordered sequentially with respect to the wavelength, the data content in the bands varies gradually as the distance between two bands grows. Therefore, when the bands in the data are ordered according to their wavelengths, the amount of correlation is usually a decreasing function of the difference in their corresponding wavelength, and hence, an image band has the highest correlation with the adjacent band. For a spectrally ordered dataset, the conditional entropy (the measure of additional information) monotonically increases with the spectral distance.

In this case, while evaluating a band I_r for its possible selection for fusion using the aforementioned scheme, the minimum value of the conditional entropy is found to be in the band from the subset $\bar{\mathbf{I}}$ whose spectral distance is the least from the band I_r. If $(p-1)$ image bands have already been selected for fusion from the first $(r-1)$ input image bands (where $p \leq r$), then the most recent $(p-1)$-th band has the least spectral distance from the remaining $(K-r)$ bands in $\mathbf{I}$.

Thus, instead of evaluating the conditional information of band I_r against all the $(p-1)$ bands of $\bar{\mathbf{I}}$ as given in Eq. (4.2), we need to compare the value for the $(p-1)$-th band only. When $\min_k H(I_r|\bar{I}_k) = H(I_r|I_k)$, Eq. (4.2) reduces to the special case as proposed in [87].

$$p = \arg \min_r \left\{ H(I_r|I_k) \geq \Theta, \ r > k \right\}. \tag{4.4}$$

The calculation of threshold Θ remains the same as given in Eq. (4.3).

The basic scheme of the entropy-based band selection process described in the previous subsection involves computation of conditional entropy of every band given $p, p = 1, 2, \ldots, \bar{K}$ selected bands, where the number of selected bands p, is monotonically increasing. On the other hand, in this special case of band selection scheme, the conditional entropy of every band is calculated only once, i.e., with respect to the most recently selected band for fusion. Therefore, the band selection process turns out to be computationally very efficient, and is very suitable for fast visualization of hyperspectral data.

4.3.1 Computational Savings

In the special case of spectrally ordered bands, a band is selected if the entropy of the band conditioned on the most recently selected band exceeds a threshold. The number of bands being selected and the corresponding average computational requirements depend on the nature of function representing the conditional information $H(I_r|I_k)$ of the image bands. When the band under evaluation is exactly the same as the band chosen, i.e., $I_r = I_k$, then the amount of additional information possessed by I_r is zero. On the contrary, when the bands I_r and I_k are totally uncorrelated from each other, then the conditional entropy of the band I_r equals its entropy, i.e., $H(I_r|I_k) = H(I_r)$. We analyze the savings in computational requirements on the basis of appropriately modeling the conditional information.

Generally, the correlation between image bands decreases exponentially as the spectral distance between the corresponding bands increases, when we may use the following theorem to compute savings in the special case of band selection scheme.

Theorem 4.1. *If the average conditional information $H(I_r|I_k)$ follows an exponential function with respect to the spectral distance $|r - k|$ with the rate parameter λ_R, then the average saving in the computation is given by $\mathscr{S} = \frac{1}{1 - \frac{\lambda_R}{\ln(1-\kappa)}}$.*

Proof: Let the conditional information be modeled as,

$$H(I_r|I_k) = H(I_r)\left(1 - e^{-\lambda_R(r-k)}\right), \quad r = k, k+1, k+2, \ldots \quad (4.5)$$

To obtain the computational savings, we use the expressions from Eqs. (4.3)–(4.4) of the band selection process.

$$H(I_r|I_k) = H(I_r)(1 - e^{-\lambda_R(r-k)}) \geq \kappa\, H(I_r),$$

$$\text{or,} \quad 1 - e^{-\lambda_R(r-k)} \geq \kappa$$

$$\text{or,} \quad (r - k) \geq \frac{1}{\lambda_R} \ln \frac{1}{1 - \kappa}. \quad (4.6)$$

Thus, a band is selected if its spectral distance from the reference band exceeds the RHS of Eq. (4.6). As the spectral distance increases, one selects lesser number of bands, resulting in higher savings in computation. When a new band is selected after discarding $(r - k)$ bands, the fractional savings $(\mathscr{S})$ in computation can be calculated as $\mathscr{S} = \frac{(r-k)}{(r-k+1)}$. In terms of the threshold κ, we can write,

$$\mathscr{S} = \frac{\left(\frac{1}{\lambda_R} \ln \frac{1}{1-\kappa}\right)}{\left(\frac{1}{\lambda_R} \ln \frac{1}{1-\kappa}\right) + 1},$$

$$\mathscr{S} = \frac{1}{1 - \frac{\lambda_R}{\ln(1-\kappa)}}. \quad (4.7)$$

The value of λ_R is dependent on the statistics of the spectrally ordered hyperspectral data. A higher value of λ_R implies a highly decreasing nature of the conditional information. Thus, the computational saving is directly proportional to the rate parameter. Also, for very small values of κ, there is practically no saving in computation, while higher values of κ lead to high values of the denominator of the fractional term in the expression, indicating a high amount of saving at the cost of selecting a very few image bands, thus sacrificing in the quality of the fusion results. As the extreme case, when κ is set to zero, the amount of savings equals zero as the system always selects every possible band in the data. On the other hand, the savings turn out to be 1 (i.e., 100%) when κ equals one when no other band gets selected.

The expression in Eq. (4.7) gives the theoretical upper bound on the computational savings. This expression, however, does not consider the processing overhead of the calculation of the conditional information. For most of the fusion techniques that operate on a per pixel basis, the amount of processing for the computation of entropy is much less as compared to the processing needed for the computation of fusion weights. The computational savings, thus, are somewhat lesser in practice.

The above discussed case of spectrally ordered data and the corresponding savings in computation assume the perfect modeling of the conditional entropy as a function of the spectral distance between the bands. In practice, there might be smaller deviations from the exponential model, which act as an additive noise. Now we analyze this more interesting case of a spectrally ordered hyperspectral data by modeling the average conditional information by an additional term that corresponds to the perturbation by an additive noise having a uniform distribution. This is useful when the model defined by Eq. (4.5) is considered partly erroneous. Since the term entropy involves an expectation operator, the corresponding quantity is a deterministic variable. We remove the expectation operator from H, and call it average information. For a given realization of the image, H may now be treated as a random variable.

Theorem 4.2. *If the average conditional information $H(I_r|I_k)$ as defined in Theorem (4.1) includes a perturbation by a white additive noise uniformly distributed in $[0, \Delta]$, then the probability of selecting the band r after having selected the band k (with $r > k$) is given by* $\min\left\{1 - \frac{H(I_r)}{\Delta}(\kappa + e^{-\lambda_R(r-k)} - 1), 1\right\}$.

Proof: As defined, $H(I_r|I_k)$ is given by,

$$H(I_r|I_k) = H(I_r)\left(1 - e^{-\lambda_R(r-k)}\right) + z \tag{4.8}$$

where $z \sim \mathbb{U}[0, \Delta]$, and typically $\Delta \ll H(I_r)$. Substituting the band selection criteria, we get,

$$H(I_r)\left(1 - e^{-\lambda_R(r-k)}\right) + z \geq \kappa\, H(I_r)$$
$$\text{or,} \quad H(I_r)\left((1 - \kappa) - e^{-\lambda_R(r-k)}\right) + z \geq 0$$
$$\text{or,} \quad z \geq H(I_r)\left(\kappa + e^{-\lambda_R(r-k)} - 1\right). \tag{4.9}$$

We denote $\upsilon = H(I_r)\left(\kappa + e^{-\lambda_R(r-k)} - 1\right)$. Then the probability of selecting the p-th band is given by,

Prob.(band r is selected, given k is the reference band)

$$= \frac{1}{\Delta}\int_{\upsilon}^{\Delta} dz \;\; = \min\left\{\frac{1}{\Delta}(\Delta - \upsilon), 1\right\}$$
$$= \min\left\{1 - \frac{H(I_r)}{\Delta}(\kappa + e^{-\lambda_R(r-k)} - 1), 1\right\}. \tag{4.10}$$

The following corollaries may be deduced from Theorem (4.2).

Corollary 1: In the limit $\Delta \to 0$, Theorem (4.1) becomes a special case of Theorem (4.2).

Corollary 2: From Eq. (4.10), $\max(r - k) = \frac{1}{\lambda_R} \ln \frac{1}{1-\kappa}$, is the same as that in Eq. (4.6). Hence, the maximum achievable savings factor in this case is also the same, i.e. $\mathcal{S} = \frac{1}{1 - \left(\frac{\lambda_R}{\ln(1-\kappa)} \right)}$. But Eq. (4.10) does not tell us how tight is the bound.

Corollary 3: If $H(I_r)$ is high, it can tolerate a proportionally larger perturbation Δ.

4.4 Experimental Results

We shall substantiate the effectiveness of the band selection scheme over the two datasets- the urban data, and the moffett$_2$ data. We shall demonstrate the results using the bilateral filtering-based fusion technique discussed in Chap. 3 as the readers are conversant with the same.

The results show that in both datasets the fused images over the subsets of both of the test data, selected using the conditional entropy-based technique are comparable in quality to the resultant images generated over the fusion of entire dataset using the same fusion technique.

Since the available bands in hyperspectral data are ordered according to wavelength, we can employ the specific case of the entropy-based selection scheme, and select the bands using Eq. (4.4). Figure 4.2a shows the result of fusion for the Hyperion data using the bilateral filtering-based fusion technique where only 27 selected image bands have been fused. This figure may be compared to Fig. 4.2b which represents the fusion of the entire hyperspectral data cube. It can be seen

Fig. 4.2 Results of fusion of the Hyperion data applied over a subset of bands selected using the conditional entropy-based approach. κ was set to 0.50 and only 27 bands were selected. **a** shows the results of bilateral filtering-based fusion over the selected bands. **b** shows the corresponding result of fusion of the entire data (© ACM 2010, Ref: [87])

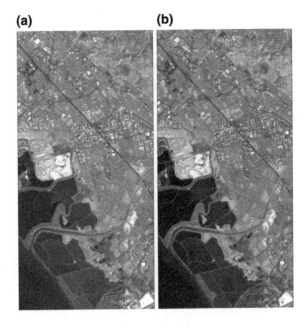

(a)   **(b)**

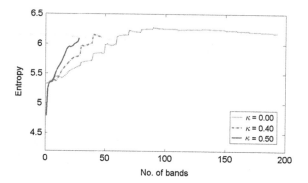

Fig. 4.3 Performance evaluation of the input-based band selection scheme as a function of the number of bands required for effective fusion of the urban data for different values of κ using the bilateral filtering-based fusion and entropy as the performance measure (© ACM 2010, Ref: [87])

that Fig. 4.2a retains most of the image features, and provides a visually comparable image quality.

We shall introduce the readers to some of the commonly used statistical parameters used for the performance evaluation of fusion results. Chapter 9 is dedicated to the performance evaluation of the hyperspectral image fusion techniques which describes various measures in detail. In this chapter, we brief readers with a few simple parameters in order to provide the performance evaluation of the entropy-based band selection scheme. Figure 4.3 represents the entropies of the fused images as more and more bands are fused progressively, for various values of κ, where the plot for $\kappa = 0$ corresponds to the fusion of the entire data. For the fusion of selected 27 image bands, corresponding to $\kappa = 0.50$, the entropy of the resultant fused image rises very rapidly, as compared to the fusion of the entire dataset. Thus, one can obtain fusion results with a very little sacrifice in the quality when compared against the resultant image from fusion of entire dataset. Also, as only a small fraction of the entire dataset is undergoing actual fusion, the fused images can be obtained in a very short time. The number of bands selected determines the savings in processing time and computation. This number is related to the value of threshold parameter κ, and thus, as κ increases there is further reduction in the computation. Therefore, we may relate Fig. 4.3 to the savings in computation. It may be noted here that as these data are spectrally ordered, we assume the time to compute $H(I_r | I_k)$ is negligible compared to the fusion process, which is often true due to the common assumption of spatial memorylessness of individual bands while computing the entropy. Figure 4.4a shows the entropies of the fused images for different values of the threshold parameter κ. It can be seen that for both test data, the performance drops very slowly beyond a value of 0.30 for κ. This indicates presence of a large amount of redundant data. Therefore, there exists an opportunity to speed up the fusion process. Another performance measure based on the sharpness of the fused image F of size (X, Y) is defined as:

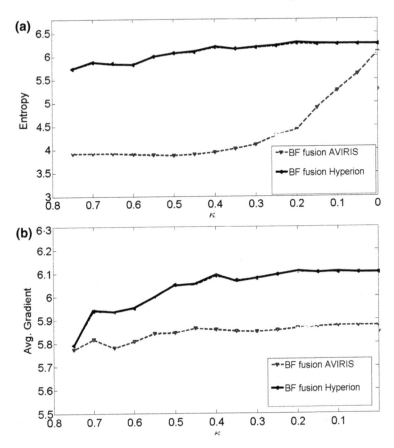

Fig. 4.4 Performance evaluation of the scheme of input-based band selection for various values of κ. **a** entropy measure, and **b** average gradient

$$\bar{g} = \frac{1}{XY} \sum_x \sum_y \sqrt{F_x^2 + F_y^2}, \qquad (4.11)$$

where F_x is the gradient of F along the x-axis. This quantity $\bar{g}$ is called *average gradient* which evaluates the quality of the image by measuring its gradient at all pixels. We can make a similar observation from Fig. 4.4b as in Fig. 4.4a which shows the variation in the values of average gradient for various values of the threshold parameter κ.

We shall compare the resultant fused images for various values of the threshold parameter κ to examine the quality of the fused images, and thus, to examine the efficiency of the band selection technique. Consider fused images for various values of the threshold parameter κ, as well as the same resulting from fusion of the entire dataset (without any band selection) when $\kappa = 0$ using the same fusion method. The result with $\kappa = 0$ is considered as the reference image and the radiometric

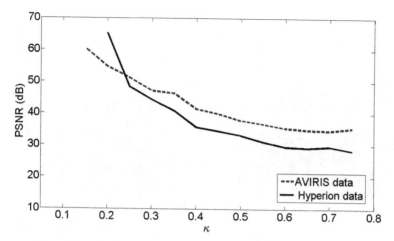

Fig. 4.5 Plots of PSNR of the fused images for various values of κ on the Hyperion and the AVIRIS data using the bilateral filtering-based fusion technique over the subsets selected using the input-based approach. The resultant image from the fusion of the entire dataset is used as a reference in each case for comparing the PSNR value (© ACM 2010, Ref: [87])

error for the fused image with $0 < \kappa < 1$ is calculated in terms of PSNR values. The nature of the PSNR values of the fused images over the subset of bands selected using the entropy-based technique can be observed in Fig. 4.5. The fused images have been obtained using the bilateral filtering-based fusion technique over the subsets of bands for various values of the threshold parameter κ. We assume the resultant image obtained through fusion of the entire dataset using the same fusion technique as the reference. As all the image bands of the moffett$_2$ data were chosen for fusion for the values of κ up to 0.15, the fusion results obtained are exactly same as the reference image. The PSNR values decrease as κ increases indicating fewer bands are being selected. It can be seen that the resultant image with the fusion of 48 bands ($\kappa = 0.40$) gives visually almost similar results with PSNR being above 40 dB. This result, in fact, has been formed from fusion of merely $1/4$-th of the original set of hyperspectral bands. A similar nature of PSNR plot can be observed for the urban data in Fig. 4.5, where a bilateral filtering-based technique has again been used for fusion. A resultant image obtained from the fusion of a subset of 69 bands chosen by selecting κ to be 0.45 was found to provide a PSNR of nearly 35 dB. For the values of κ less than 0.20, the band selection scheme selects all the image bands in the Hyperion dataset, and produces the same final result of fusion as the reference image. Thus, for both datasets, the band selection technique proves to be capable of selecting small, yet efficient subset of hyperspectral bands that can produce visually almost comparable results even when a small fraction of the original data is selected.

A visual comparison of the fusion results of the entropy-based band selection can be carried out from Fig. 4.6 where the resultant images obtained from the fusion of selected subsets of the AVIRIS data using the bilateral filtering-based fusion technique are shown. The images in Fig. 4.6a–d are obtained from the subsets of

(a)   **(b)**

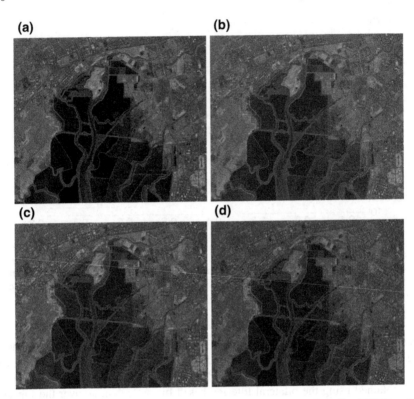

(c)   **(d)**

Fig. 4.6 Results of bilateral filtering-based fusion of the moffett$_2$ data from the AVIRIS applied over a subset of bands selected using the entropy-based approach. **a–d** show the results of fusion for κ values 0, 0.35, 0.50 and 0.70, respectively (Fig. 4.6c: © ACM 2010, Ref: [87])

the moffett$_2$ data of different cardinality by varying the threshold parameter κ. We consider the resultant image from fusion of the entire dataset as shown in Fig. 4.6a as the reference for the quality assessment purpose. The result of fusion of 65 selected bands of the data as shown in Fig. 4.6b reduces the computations nearly to the one-third of the original. However, the corresponding fusion result does not produce any visible artifacts, and gives a PSNR value of 41 dB when compared against the aforementioned reference. The result of fusion of 29 selected bands was found to give 35 dB of PSNR (Fig. 4.6c). A set of 12 image bands was selected by choosing κ to be 0.70. The resultant fused image is shown in Fig. 4.6d which was found to give a PSNR of 30 dB for the fusion of approximately $1/10$-th of the data.

Along with visual analysis, we also need to evaluate the quality of the fused images obtained from the selected subsets with respect to the fused image obtained from the entire dataset in terms of their probability distribution functions (pdfs). We measure the extent of the similarity between the histograms of the corresponding two images. We provide this analysis using the Bhattacharyya distance which measures the overlap between the histograms of the two images. As in the previous case,

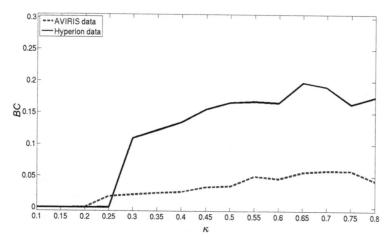

Fig. 4.7 Plots of Bhattacharyya coefficient between the resultant image from the fusion of entire data and the resultant images for various values of κ for the urban and the moffett$_2$ data using the bilateral filtering-based fusion technique over the subsets selected using the input-based band selection approach (© ACM 2010, Ref: [87])

the resultant image from fusion of the entire dataset is considered as the reference. Figure 4.7 provides the plots of the Bhattacharyya coefficient (BC) between the reference image and the resultant images obtained from the fusion of a subset of image bands obtained for different values of κ, selected using the entropy-based scheme for the urban data as well as the moffett$_2$ data. The images have been fused using the bilateral filtering-based technique. The Bhattacharyya coefficient (BC) can be observed to steadily increase for increasing values of κ. This behavior indicates gradual deviation (in terms of histogram overlap) of the fused images from the reference image as lesser number of bands get selected for fusion. The entire set of hyperspectral images was selected for very small values of κ, when the BC is zero, as it can be observed from both plots. It may be observed from Figs. 4.4, 4.5 and 4.7 that the performance of the band-selection method is very much data dependent, i.e., AVIRIS and Hyperion data having different sensor characteristics. However, it may be observed from the nature of the plots that the band selection scheme performs equally well on both datasets.

In the case of spectrally ordered hyperspectral data, the band selection scheme reduces a significant amount of computational time, however it has an overhead of the calculation of the conditional entropy of each image band. The band selection scheme is beneficial only when the computation of actual fusion algorithm far exceeds the computation of the entropy of each of the image band. However, except trivial and simple techniques like averaging, for most of the existing robust fusion techniques the time needed for band selection is much smaller than the time taken for fusion of the entire dataset. Especially when the fusion techniques operate on a per pixel basis, or they involve some kind of iterative routines, the computation of entropy is quite negligible as compared to the computational requirements of actual fusion

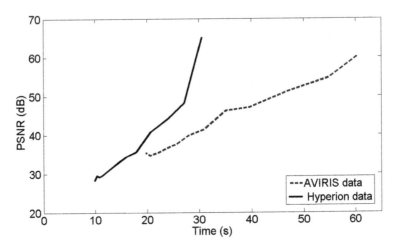

Fig. 4.8 Plots of PSNR of the fused images against the timing requirements for the urban and the moffett$_2$ data for the bilateral filtering-based fusion technique. The resultant image from the fusion of the entire dataset is used as a reference in each case for the evaluation of the PSNR (© ACM 2010, Ref: [87])

process. Therefore, the entropy-based subset selection scheme proves to be highly effective for complex and computationally extensive techniques of fusion. The total computation $\mathbf{W}$, taken for the pixel-level fusion procedure as a function of threshold κ, is of the form-

$$\mathbf{W}(\kappa) = \gamma \,\mathscr{B}(\kappa) + c_E \qquad (4.12)$$

where $\mathscr{B}(\kappa)$ represents the number of bands selected for a given threshold κ, and c_E is the amount of computation for the evaluation of conditional entropies of the image bands for the band selection procedure. This second term is a constant for a given dataset. The γ factor is a proportionality factor to account for computational requirements of a given fusion technique. If the time required for the computation of the conditional entropy is negligible as compared to the timing requirements of actual fusion, we can approximate Eq. (4.12) as, $\mathbf{W}(\kappa) \approx \gamma \,\mathscr{B}(\kappa)$ which indicates that the order of computation is linear with respect to the number of bands selected. It should be noted that as κ increases, $\mathscr{B}(\kappa)$ decreases leading to a lesser amount of computation, however this does not affect the linear nature of the computational complexity.

For a qualitative analysis of the band selection scheme for the computational requirements, we provide the plot of PSNR of fused images against the total time taken by varying the threshold parameter κ. Figure 4.8 depicts the plots for the urban and the moffett$_2$ datasets for fusion using the bilateral filtering–based technique. A monotonically increasing nature of these plots indicate the improvement in the quality of the fusion result at the expense of computational requirements when more bands are fused. The reason that the computational requirements are nearly double for the AVIRIS data compared to the Hyperion data lies in the fact that the image size for the AVIRIS is much larger in our study.

4.5 Summary

This chapter discusses an information theoretic scheme for the selection of specific image bands for fast visualization of hyperspectral data. The conditional entropy-based selection scheme selects a subset of bands from the input which are mutually less correlated. Thus, this subset retains most of the information contents in the data, and assures a minimal degradation in the quality of the result of the fusion. As only a fraction of the entire dataset undergoes fusion, the band-selection approach facilitates a faster and computationally efficient fusion. The correlation pattern among the adjacent bands in a spectrally ordered hyperspectral data enables us to accelerate the process of band selection. A band is evaluated for the purpose of fusion by calculating its additional information content against only the last selected band.

A performance evaluation of the band-selection technique in terms of the quality of the subsequent results of fusion using the bilateral filtering-based fusion technique has also been presented in support of the efficiency of the band-selection technique.

Chapter 5
Bayesian Estimation

5.1 Introduction

In Chap. 3, we addressed the problem of visualization of hyperspectral data through a pixel-level fusion of the constituent bands. The fused image has been generated as the composite of the input spectral bands where the locally dominant features in the data were used to determine how the pixels should be combined. For pixel-level methods, the fusion weights are typically calculated from some measure that describes the relative importance of the pixel (or the band) in the data with reference to its spatial neighborhood. This importance measure, also known as *saliency* quantifies how important the pixel is for fusion purpose. The fusion technique discussed in Chap. 3, makes use of the bilateral filter for the extraction of details from each and every pixel in the hyperspectral data.

In this chapter, we shall explore a different aspect of fusion process. Let us take a look at the process of image formation. We assume that the fused image represents the *true* scene. The elements of the hyperspectral sensor *see* the same scene, however, capture it partially across the set of bands due to the specific spectral responses of the sensor elements. Thus, each of the hyperspectral bands captures some fraction of the true scene. The objective of the fusion process is then to obtain the resultant image as close to the true scene through combining all available input bands representing the scene partially. The first step towards the solution is to formulate the relation between the true scene (to be obtained through fusion), and each of the constituent bands using a certain model. One may consider the statistical model of image formation which relates the true scene and input images using a first order approximation. According to this model, every sensor element captures only a fraction of the underlying true scene. This fraction is regarded as the sensor selectivity factor in this model, which primarily reflects how well the particular pixel has been captured by the corresponding sensor element of the particular band of the hyperspectral image. When the sensor selectivity factor is 1, the corresponding pixel in the true scene has been seen exactly by the corresponding sensor. On the other hand, if the sensor selectivity factor is zero, the

S. Chaudhuri and K. Kotwal, *Hyperspectral Image Fusion*,
DOI: 10.1007/978-1-4614-7470-8_5, © Springer Science+Business Media New York 2013

pixel in the image band has no correlation with the true scene, and hence, it can be regarded as noise.

The problem of fusion is now equivalent to the estimation of parameters of the image formation model. Ideally, one can compute these parameters from the precise information about the sensory system. However, this information is not always available. Also, the method does not remain generic after incorporation of the device-specific information. Therefore, let us assume no knowledge about the sensor device, and address this problem blindly using only the available hyperspectral data. The parameters of the image formation model should be computed from certain properties of the data. The model, however, being a first order approximation does not reflect a perfect relationship between the input and the output images. The modeling error may lead to an erroneous resultant image. Considering this as random perturbations and to deal with these perturbations in an efficient and systematic manner, we formulate the estimation problem in a Bayesian framework. The solution is obtained through the maximum *a posteriori* estimate of the corresponding formulation.

In the next section we explain in brief the Bayesian framework and some of its applications in vision and image processing. We describe the image formation model developed by Sharma et al. [164] in Sect. 5.3. The computation of the parameters of this model is explained in Sect. 5.4. The Bayesian solution and its implementation details are provided in Sects. 5.5 and 5.6, respectively. Some illustrative results are provided in Sect. 5.7, while Sect. 5.8 summarizes the chapter.

5.2 Bayesian Framework

We often come across with problems in various fields of engineering where the data available to us are either incomplete or imperfect. The data, which might be available in the form of a set of observations from some system, may contain additional components, unwanted for the given application. When one wants to make some inferences or calculate some parameters from such data, results are likely to be erroneous due to the unwanted component, known as noise. Alternatively, one may not have an exact model of the system under investigation, which can also lead to imprecise solutions. Thus, the uncertainty or the incompleteness of the data or model are primary sources of errors in the inference.

Bayesian estimation is a probabilistic framework that has been extensively used for reasoning and decision making under uncertainty. It provides a systematic way of approaching the solution when the knowledge about the system is incomplete. This framework takes into account the posterior probability related to a specific event pertaining to the quantity of interest. A posterior probability refers to the conditional probability assigned after the relevant event has been taken into consideration. Bayesian estimator provides a solution by maximizing the posterior probability. Due to incomplete knowledge, one may not be able to obtain the exact solution, but can obtain a solution that is close to the desired one. The process of Bayesian estimation can be illustrated as follows. Let θ be some underlying variable of interest while we have been given the observation ζ. We would like to estimate θ, i.e., to find the best possible value of θ from the given observation ζ. Firstly, we want to compute the posterior probability of θ given the observation ζ as in Eq. (5.1).

$$\mathscr{P}(\theta|\zeta) \propto \mathscr{P}(\zeta|\theta)\,\mathscr{P}(\theta). \tag{5.1}$$

The first term in this expression refers to the conditional probability of the occurrence of the observation ζ given θ which is the underlying variable of interest. The term $\mathscr{P}(\zeta|\theta)$ is known as the likelihood of θ. The likelihood functions play an important role in Bayesian estimation.

The second term $\mathscr{P}(\theta)$ in Eq. (5.1), is the probability density function (pdf) of the variable of interest, θ. Although the exact or precise information about the θ variable is not available, some knowledge about the same is generally available. It could be obtained from the physical nature of the system, history, or some known constraints on the system. This information reduces some amount of uncertainty associated with θ, and hence it is also referred to as the prior probability, or simply the prior.

The prior information $\mathscr{P}(\theta)$ may not always be available. Depending upon whether the prior information is available or not, a Bayesian framework can be categorized into the following two estimation methods-

1. Maximum likelihood (ML).
2. Maximum *a posteriori* (MAP).

The ML estimator neglects the prior probability of the quantity of interest, and maximizes only the likelihood function, i.e., the probability defined by the model. The principle of the maximum likelihood (ML) estimator is to find the value of variable θ that maximizes the likelihood function, i.e.,

$$\hat{\theta}_{ML} = \arg\max_{\theta} \mathscr{P}(\zeta|\theta). \tag{5.2}$$

It is a common practice to deal with the logarithmic value of the likelihood function due to mathematical convenience. However, it does not alter the solution as the log is a monotonically increasing function. One of the most commonly used examples of the ML estimation is the least squares method, when the perturbations are Gaussian distributed.

The maximum *a posteriori* (MAP) estimation is closely related to the ML estimation, but incorporates the prior distribution of the quantity to be estimated. Consider the expression for the ML estimation given by Eq. (5.2). The expression for MAP is obtained by augmenting the objective with the prior $\mathscr{P}(\theta)$. The MAP estimation, thus, refers to estimation of the value of θ that maximizes the expression in Eq. (5.1). The MAP estimation can be formulated by Eq. (5.3).

$$\hat{\theta}_{MAP} = \arg\max_{\theta} \mathscr{P}(\zeta|\theta)\mathscr{P}(\theta). \tag{5.3}$$

Similar to the ML estimation procedure, the log-likelihood function is often used in the MAP estimation for its computational ease without affecting the results. The connection between the two techniques of estimation is obvious. The MAP estimate equals the ML estimate when the prior $\mathscr{P}(\theta)$ is uniform, i.e., a constant function. The MAP estimate can also be considered as a regularized ML estimate. Additionally, when the prior is non-informative, then the MAP estimate approaches the ML estimate, i.e, $\hat{\theta}_{MAP} \to \hat{\theta}_{ML}$.

The ML and MAP estimators provide simple, but powerful frameworks of Bayesian statistics to deal with a variety of problems through the right choice of the model and the prior. Some highly popular techniques such as Viterbi algorithm, Kalman filter, hidden Markov model, and Expectation-Maximization algorithm have their roots in the Bayesian framework. While these techniques have been used to solve problems from various fields, here we discuss in brief a few techniques and applications relevant to image processing and vision.

Geman and Geman introduced a Bayesian paradigm for the processing of images using the Markov random field (MRF) [63]. They demonstrated the results in connection with the image restoration problem which searches for the MAP estimate of an image modeled as a Markov random field using simulated annealing. Their method has been further explored by several researchers. The assessment of MAP-based restoration for binary (two-color) images has been carried out by Greig et al. where the performance of the simulated annealing has also been tested [70]. Murray et al. have experimented with the MAP restoration of images on the parallel SIMD processor array for speeding up the operation [120]. The problem of reconstruction of medical images has often been addressed through the Bayesian framework. An iterative algorithm for the reconstruction of PET images using a normal prior has been proposed in [73]. This solution has a guaranteed convergence which is optimal in the MAP sense. A 3-D reconstruction of images obtained by the microPET scanner using the MAP formulation has been discussed in [143], where the likelihood function and image have been modeled using the Poisson and the Gibbs priors, respectively. Lee et al. have incorporated a weak plate-based prior for the reconstruction of tomography images, which is claimed to preserve edges during the reconstruction [100].

Several edge preserving operations have the Bayesian concept at their basics. The anisotropic diffusion proposed by Perona and Malik [130] can be formulated into a Bayesian framework [11]. The bilateral filter discussed in Chap. 3 also has a Bayesian connection as shown by Barash [11, 12]. A classical problem of image segmentation has also been formulated into a Bayesian framework. Sifakis et al. have proposed a Bayesian level-set formulation for image segmentation [166]. The use of the double Markov random field [112], and the binary Markov random field [48] has also been investigated for the image segmentation problem. Moghaddam et al. have developed a MAP-based technique for visual matching of images which can be used for face recognition or image retrieval [119]. In [18], a MAP-based object recognition technique has been proposed which employs an MRF-based prior to capture phenomena such as occlusion of object features.

We have already discussed the use of a Bayesian framework for image fusion in Chap. 2. Sharma et al. have modeled the image observations as affine transformations of the underlying true scene [164]. The true scene is obtained using a MAP estimator. Yang and Blum have proposed modifications to this model in a multi-scale decomposed framework [196]. Kumar [94], and Kumar and Dass [95] have also demonstrated effectiveness of the Bayesian technique for generalized image fusion. For the problem of fusion, the images to be fused act as observations, and the fused image is the quantity of interest. These two are related to each other using an appropriate image formation model which defines the corresponding likelihood function. Now we shall explain in detail the image formation model.

5.3 Model of Image Formation

In the past decade, Sharma et al. introduced a statistical technique for generalized image fusion. Their technique consists of defining an image formation model, the perturbation noise modeled with a Gaussian distribution, and a Bayesian technique to solve the fusion problem [164]. This model of image formation is given by Eq. (5.4).

$$I_k(x, y) = \beta_k(x, y)\, F(x, y) + \eta_k(x, y), \tag{5.4}$$

where $I_k(x, y)$ denotes the observation of a true scene pixel $F(x, y)$ captured by the sensor k. $\eta_k(x, y)$ indicates the noise or disturbance component. $\beta_k(x, y)$ is known as the sensor selectivity factor which determines how well the given observation has captured the true scene. The maximum value that $\beta_k(x, y)$ can achieve is unity, which indicates that the particular pixel is exactly same as it should be in the fused image (in the absence of noise). On the other extreme, $\beta_k(x, y)$ can have a value of zero, when the pixel is essentially pure noise without any contribution towards the final result.

Yang and Blum have proposed a multiscale transform version of this fusion technique that can also handle non-Gaussian disturbances [196]. In their model the sensor selectivity factor can assume values of 0 or ± 1 only. The negative value indicates a particular case of polarity reversal for IR images [196]. This discrete set of values brings out only two possibilities- either the sensor can *see* the object, or it fails to *see* it. The β factor needs to be calculated from the available data only. In [196], the value of β has been calculated by minimizing a function related to the sensor noise.

Fusion using the statistical model of image formation has been enhanced by allowing the sensor selectivity factor β to take continuous values in the range [0, 1] in [94, 95]. However, the value of β has been assumed to be constant over smaller blocks of an image [95, 164]. These values of β have been computed from the principal eigenvector of these smaller blocks. The values of β, therefore, are constant over image blocks, but can be totally discontinuous across the adjacent blocks due to the data dependent nature of eigenvectors. Xu et al. have modeled the sensor selectivity factor using a Markov random field (MRF), however it can take values from the discrete set {0, 1} only [192, 193]. They have modeled the fused image also to be an MRF which acts as the prior for the MAP formulation.

Ideally, the values of the sensor selectivity factor should be continuous within the range [0, 1]. Every pixel corresponds to a different spatial location in the scene with a similar but different material composition. Therefore, every pixel should have a different value of sensor selectivity factor β. We can, however, expect these values to exhibit some degree of spatial correlation within a given band. In the next section we explain a solution to compute the sensor selectivity factor β which conforms to the aforementioned constraints.

5.4 Computation of Model Parameters

Through the sensor selectivity factor β, one is primarily interested in mapping the quality of the corresponding pixel in the input hyperspectral data. Yang et al. have associated the sensor selectivity factor β with the ability of a given sensor to *see* the objects [196]. While Kumar et al. consider β to be the gain of the sensor [95]. With this analogy, β can assume any values in the range [0, 1]. Their computation of β is based on the eigenvectors of small image blocks which makes the β-surface constant over these image blocks. However, this β-surface can be totally discontinuous across adjacent blocks in the given band. Also, if this technique is to be implemented over hyperspectral data, it will prove to be computationally very demanding due to a huge volume of the data.

We associate the value of β to the perceived quality of the pixel from a particular band. As the primary goal of fusion is visualization, we want pixels with higher *visual quality* to contribute more towards the final fused image. The conventional fusion weight defines the contribution of the pixel relative to rest of the observations from the entire set of input bands. In the present case, we are dealing with a single band and the fused image using the image formation model. Although the visual or perceived quality is best judged by the human observer, employing subjective measures is not a viable and practical solution. One can, however, employ various objective quality measures that can be closely related to the visual quality of the pixel. The objective measures offer a uniform and repetitive assessment of the pixel quality. Several no-reference quality measures have been discussed in the literature. These no-reference measures can be easily calculated from the image itself without any reference or a ground truth. The use of these measures eliminates the dependency on any other standard or reference data, and hence, makes the quality assessment a stand alone process. Also, in most cases, the quality measures such as the contrast or sharpness, are easy to calculate, and computationally efficient. In this section we explain the use of no-reference quality measures locally at every point to compute β. While a certain single quality measure of the image can definitely be used for this purpose, the use of only one measure may not be able to efficiently quantify the pixel quality, and thereby the sensor selectivity factor β. Different quality measures capture different aspects of the *visual quality* of the image, and thus, we need a combination of such measures that capture complementary aspects of the pixel quality. We develop a simple, yet efficient way to compute the sensor selectivity factor from a set of pre-defined quality measures of the input data. We calculate the sensor selectivity factor $\beta_k(x, y)$ for a particular pixel (x, y) in a particular band k as the product of different individual quality measures of the image. If $Q_1, Q_2, \cdots, Q_n$ are the different quality measures, then the value of β is given by Eq. (5.5).

$$\beta_k(x, y) = \prod_{j=1}^{n} Q_j(I_k(x, y)), \tag{5.5}$$

where $Q_j(.)$ indicates evaluation of the j-th quality measure. These quality measures, however, should satisfy certain properties. Though we are not dealing with the conventional fusion weights, the non-negativity of the sensor selectivity is an essential condition. This factor reflects the fractional component of the true scene captured by the given sensor element. Intuitively, and by definition, this component has to be non-negative. The minimum value of β can be zero, which indicates pure noisy data without any relation to the true scene. The quality measure should, therefore, produce a non-negative value, and it should also be confined to a finite range. If the quality measure produces a negative value due to reasons such as small magnitude, phase reversal, etc., the multiplicative combination of measures given in Eq. (5.5) may as well turn out to be negative, and change the notion of the corresponding measure. Non-negative values enable a direct combination of multiple quantities without a need for sign change. The value of β is calculated as the product of the local quality measures, and hence, even if a single measure has the value of zero, the corresponding $\beta_k(x, y)$ will be enforced to be zero producing erroneous results. Secondly, these measures should be confined to a finite range. We do not need a normalization across all the bands of hyperspectral image, as typically done for the conventional fusion weights. However, the β values should be confined to a specific range. The designated range for β is $[0, 1]$, however the actual value of products of chosen quality measures may not confine to this range. One has to appropriately scale the computed product of quality measures to the designated $[0, 1]$ range. We have to compute these quality measures over each pixel in each spectral band of the hyperspectral data. Therefore, it is desirable to select computationally efficient quality measures Q_j's.

Out of the various existing no-reference quality measures, we choose the following two:

- well-exposedness, and
- sharpness.

The gain of the hyperspectral sensor array as a function of the wavelength varies greatly over the entire span of the bandwidth of spectral bands. Pixels with very high values of intensity appear too bright due to over-saturation. On the other extreme, poorly-exposed or under-saturated pixels appear too dark, and thus, fail to convey visually useful information. As these pixels do not provide any significant information, one naturally expects lesser contribution from such pixels towards the final fused image. The sensor selectivity factor for such pixels should obviously be small.

We expect this factor to be high for pixels having their intensity far from both the extremes of the dynamic range. For example, if we are working with a normalized data with grayscale ranging from 0 to 1, the β values for the pixel intensities close to 0.50 should be higher. To quantify this component, we assign weights depending upon the distance of the gray value of the pixel from the middle value of 0.50 for the normalized data. We calculate the distance of the gray value of a pixel from 0.50 over a Gaussian curve having a fixed spread parameter σ_β. The weights are, thus, inversely proportional to the squared difference in the intensity of the pixel from the mid-intensity value (0.50 in the case of normalized data). A similar quality measure has been employed for the fusion of multi-exposure images in [89, 113]. Equation (5.6) provides the evaluation of the first quality measure Q_1 over an observation $I_k(x, y)$.

$$Q_1(I_k(x, y)) \equiv \exp\left(-\frac{(I_k(x, y) - 0.50)^2}{2\sigma_\beta^2}\right). \tag{5.6}$$

The spread (or variance) parameter σ_β^2 controls the width of the Gaussian, and thus, the relative quality assigned to the pixel. This parameter can be selected depending upon the requirement of the result. For too small values of σ_β, only the pixels having gray values close to the half, will contribute towards the fusion result. As σ_β increases, more pixels get incorporated into the fusion process.

A sharp image is not only visually appealing, but it also offers several practical advantages. Sharp images facilitate easy interpretation by a human analyst. Also, a large number of machine vision algorithms can produce better results when input images have sharp features present. For example, remote sensing images contain an agglomeration of a large number of small objects, such as residential areas, roads, etc. For a better understanding and efficient further processing, these objects should be clearly observable in the fused image. Sharp edges and boundaries enable quick and accurate identification of such objects. The performance of several processing algorithms, especially the ones involving segmentation, gets significantly improved due to presence of strong and sharp features. As the sharper regions bring more visual and useful content, we expect such regions to have higher contributions towards the fused image which represents the true (ideal) scene. The sensor selectivity factor β for a sharp region in the particular band should be higher. We apply a Laplacian filter to pixels in the individual hyperspectral bands, and measure its absolute value which provides the local sharpness. The use of Laplacian filter alone yields zero output in uniform or homogeneous regions. To circumvent this problem, we define the second quality measure Q_2, by adding a small positive constant C to the output of the Laplacian filter as the following-

$$Q_2(I_k(x, y)) \equiv |\nabla^2(I_k(x, y))| + C, \tag{5.7}$$

where ∇^2 denotes the Laplacian operator and C is a small positive constant. When a pixel lies on some sharply defined objects such as edges, the first term in Eq. (5.7) is high, producing a high value of the corresponding quality measure Q_2. For pixels with small values of the gradient in their local neighborhood, this term in the expression for sharpness measure (Q_2) is quite close to the constant C, resulting in comparatively smaller values of the corresponding measure Q_2. It should be noted that the range of Q_2 depends on the choice of the constant. We need to explicitly scale Q_2 appropriately over k for each spatial location (x, y), and confine it to the range $[0, 1]$.

The constant C serves two purposes—first, it provides a non-zero value to the quality measure Q_2 in the uniform or homogeneous regions. This prevents the possibility of the sensor selectivity factor β attaining zero values solely due to the uniform nature of the neighborhood region. Secondly, the constant C brings the flexibility to the fusion system. Very few pixels from a particular band possess a significant value of Q_2 when C is small. That is, only those pixels which have a very high value of sharpness in terms of their spatial gradient will get selected. With increase in the value of C, more and more pixels get selected for the actual process of fusion. The constant acts as the fixed bias or the offset for the selection of good pixels. Thus, with the help of constant C, the user can tune the fusion algorithm as per the requirements of the output.

We can obtain the sensor selectivity factor β for every pixel of the data by substituting Eqs. (5.6) and (5.7) in Eq. (5.5). These values, however, have been calculated independently on a per pixel basis. The adjacent pixels in the hyperspectral image bands usually have the same or similar material composition, and thus, their reflectance responses are also highly similar. The degree of spatial correlation among the intra-band pixels is quite high. As the sensor selectivity factor β indicates visual quality of the pixel, one may naturally expect this factor also to possess some degree of spatial closeness. We do not expect sharply discontinuous nature of β within a local region in a given hyperspectral band. It should, however, be noted that the β values within a small neighborhood should not be exactly the same as each pixel is unique. We redefine the problem of computation of β in a regularization framework to address this requirement of spatial smoothness. We incorporate a regularizing term that enforces the spatial smoothness constraint on the sensor selectivity factor β. This additional term penalizes departure of the sensor selectivity factor β, from smoothness, and prevents sharp discontinuities within the given band. The computation of $\beta_k(x, y)$ proceeds as the minimization of the following expression given in Eq. (5.8).

$$\beta_k(x, y) = \arg\min_{\beta_k} \int_x \int_y \left(\left(\prod_{j=1}^{2} Q_j(I_k(x, y)) - \beta_k(x, y) \right)^2 \right.$$
$$\left. + \lambda_\beta \left(\beta_{k_x}^2(x, y) + \beta_{k_y}^2(x, y) \right) \right) dx\,dy, \qquad (5.8)$$

where λ_β indicates the relative importance of the smoothness constraint, also known as the regularization parameter. β_{k_x} and β_{k_y} indicate the spatial derivatives along the

x- and y-directions, respectively. The solution of such problem is typically given by the Euler-Lagrange equation. We shall provide more details on the Euler-Lagrange equation with reference to the calculus of variations in the next chapter. (Refer to Sect. 6.2). The Euler-Lagrange equation provides an iterative solution for $\beta_k(x, y)$ as given in Eq. (5.9).

$$\beta_k^{(m+1)}(x, y) = \bar{\beta}_k^{(m)}(x, y) - \frac{1}{2\lambda_\beta}\left(\beta_k^{(m)}(x, y) - (I_k(x, y))\right), \qquad (5.9)$$

where (m) is the iteration number. The values of $\beta_k(x, y)$ where the product of the corresponding quality factor terms is too close to zero or one, may lie outside the desired range of $(0,1]$ when solved using Eq. (5.9). Ideally, this can be prohibited by introducing a set of some explicit constraints to Eq. (5.8). However, this process converts the problem of β computation into a constrained optimization problem which is computationally demanding. To speed up the computation process we refrain from the use of explicit constraints. Experimentally we found that very few values of the selectivity factors lie outside the range, and they have been clipped to the corresponding extrema, i.e., 0 or 1. This enables us to solve the unconstrained optimization problem without much degradation in the quality of the results.

The hyperspectral data also exhibit a moderate-to-high correlation along the spectral array. Therefore, one may want to impose the smoothness constraint over the spectral dimension (k) as well, and obtain the sensor selectivity factor $\beta_k(x, y)$ which is globally smooth. This requires incorporation of additional terms to Eq. (5.8), which are related to the derivative of β across the spectral dimension k. However, such a process for computation of the sensor selectivity factor β will be very slow as compared to the existing one, and thus, we refrain from the use of this additional derivative term. It should, however, be noted that the penalty for departure from the smoothness is chosen to be quite weak. The purpose of introducing this penalty is to address the spatial relationship among the pixels, and prohibit arbitrary spatial variations in β. However, one should be careful in selecting the penalty factor, and avoid a strong penalty which can oversmooth the β-surface. A strong penalty factor is not desired.

The aforementioned procedure generates a β-surface for every band in the hyperspectral data. Let us consider an original band of hyperspectral image as shown in Fig. 5.1a. This band depicts a small region of Palo Alto (CA, USA), constituting the urban area. This is the 50-th band of the urban data by the Hyperion sensor. Figure 5.1b shows values of β computed using Eq. (5.8). These values, which are formed by the product of two measures associated with the visual quality of the corresponding scene pixels, attempt to quantify the usefulness of the underlying pixel towards the final result of fusion. It may be observed that higher values of β correspond to the areas that are sharp and clearly visible in the image, while smaller values of β have been obtained for dull, less informative areas.

Fig. 5.1 Illustration of the computation of sensor selectivity factors (β) for the urban image from the Hyperion. **a** Original image is the 50-th band in the urban data, and **b** computed β image for the same band

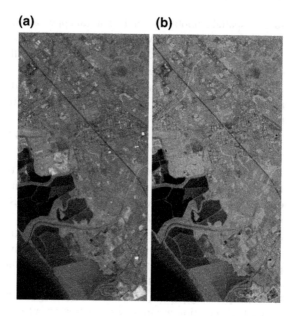

(a)　　　　**(b)**

5.5 Bayesian Solution

We would like to estimate the true scene, i.e., the fused image F from the noisy observations (bands). These observations I_k, are related to the fused image F through a first order model. A pixel in the fused image needs to be estimated from the array of observations across all hyperspectral bands at the same spatial location, (x, y). Therefore, during this process, we mostly deal with the 1-D spectral data at a given location in a 2-D spatial plane. Therefore, instead of working with 2-D bands I_k, $k = 1, 2, \cdots, K$, we work with spectral array notation as explained earlier in Chap. 2. Accordingly, the array of spectral observations at (x, y) shall be referred to as $\mathbf{s}(x, y)$ such that $\mathbf{s} \in \mathbb{R}^K$. The observation in the k-th band, i.e., the k-th element of this vector is denoted by $s_k(x, y)$. It should, however, be noted that the change of notation is purely meant for ease of understanding, and to avoid any possible confusion when the same data are being referred from different dimensions.

The maximum likelihood (ML) solution is a commonly preferred approach for estimation from a noisy data as discussed earlier. The ML solution produces an output by minimizing the quadratic norm of the error between the estimated fused pixel and the corresponding quantity related to input pixel scaled by β when the noise is assumed to be Gaussian. The ML solution estimates each pixel of the fused image F independent of the neighborhood pixels, and thus, the spatial correlation within pixels of the output image has not been taken into consideration. Most of the natural images are spatially smooth, except at the edges. This moderate smoothness is one of the important characteristics of natural images. As the ML solution operates on a per pixel basis, it may produce an image that lacks the properties such as spatial smoothness, and a natural appearance. One of the common strategies for improvements in the ML estimate is to incorporate some prior information on the image to be estimated. This process transforms the problem into the maximum *a posteriori* (MAP) estimation.

The problem of MAP-based estimation of the fused image $\hat{F}$ can now be described as obtaining the function $\hat{F}$ that maximizes the following joint probability density function-

$$\hat{F}(x, y) = \arg\max_{F} \mathscr{P}(s(x, y)|F(x, y)) \, \mathscr{P}(F(x, y)). \qquad (5.10)$$

The MAP formulation, as explained, is composed of two terms. The first term on the RHS of Eq. (5.10) represents the likelihood function of F, i.e., the conditional pdf of the hyperspectral observations $s(x, y)$ given the fused image $F(x, y)$. The second term represents the pdf of F known as the prior on F. The prior favors certain desired outputs, and thus, reduce the uncertainty in the solution.

First we consider the image formation model given by Eq. (5.4). Having computed the sensor selectivity factor $\beta_k(x, y)$, we want to estimate the fused image F from the noisy observations. The noise term $\eta_k(x, y)$ represents the departure of the corresponding pixel $s_k(x, y)$ from the image formation model. The noise term for each observation at any given location (x, y) is assumed to be an independent and identically distributed (i.i.d.) Gaussian process in [95, 164, 192, 193]. We also model the noise by a zero mean Gaussian i.i.d. process with the variance value of σ^2. We can write the conditional probability density function (pdf) of the observed spectral array s_k given the sensor selectivity factor β_k, and the fused image F using Eq. (5.11).

$$\mathscr{P}(s(x, y)|F(x, y)) = \frac{1}{(\sqrt{2\pi\sigma^2})^K} \exp\left(-\frac{||s(x, y) - \beta(x, y) F(x, y)||^2}{2\sigma^2}\right) \quad \forall (x, y). \tag{5.11}$$

We omit the function arguments x and y for the purpose of brevity (i.e., $\beta \equiv \beta(x, y)$, etc.), and write the log-likelihood function as the following.

$$\log \mathscr{P}(s|F) = \log\left(\frac{1}{(\sqrt{2\pi\sigma^2})^K} \exp\left(-\frac{||s - \beta F||^2}{2\sigma^2}\right)\right) \quad \forall (x, y)$$

$$\log \mathscr{P}(s|F) = -\frac{K}{2} \log(2\pi\sigma^2) - \frac{||s - \beta F||^2}{2\sigma^2} \quad \forall (x, y). \tag{5.12}$$

Now, since we have provided the basic MAP formulation, let us consider the second term in Eq. (5.10) which deals with the *a priori* information on the fused image F. One can derive the prior term from the available information about the function to be estimated. The prior information can also be obtained from some of the physical properties that the estimated quantity is expected to have. A Gauss-Markov random field (GMRF) is one of the popular prior in image and vision problems. In [192, 193], the fused image has been modeled as an MRF in order to obtain a spatially smooth resultant. The MRF-based model represents local behavior among the pixels in the form of the Gibbs distribution, where the value of the fused pixel is a function of pixels in the spatial neighborhood of the corresponding location. The use of the MRF-based prior typically produces spatially smooth images by favoring the radiometrically similar pixels, and penalizing sharp discontinuities. However, the GMRF priors tend to blur the edges and boundaries by over-smoothing them, unless the edge fields have been separately modeled. The corresponding output cannot be a good approximation of the image to be estimated if it contains edges [28]. Remote sensing images typically contain a large number of small objects, which can be identified quickly only if the corresponding edges and boundaries are sharp and clear. Therefore, the use of any prior that leads to blurring or smearing of edges, should be avoided. We would like to use the prior that preserves and enhances the edge information in the data. Rudin et al. proposed the use of the total variation (TV) norm based on the L_1 norm of the derivatives of image [155]. Initially proposed for image denoising, the TV-norm has also proved to be very effective in several image processing and image restoration applications [27, 51, 57, 122]. The TV norm–based penalty functions can be approximately modeled as a Laplacian pdf prior, and thus, one may use them as a prior within the MAP formulation also. The TV norm based priors preserve the discontinuities in the image, and thus, retain sharp boundaries and edges. The fused image preserves small objects and features in the scene, which would get washed away or smeared during the process of fusion otherwise. When employed in the deterministic framework, it is also claimed to provide a better quality as compared to the one obtained by the use of the L_2 norm-based penalty functions under exactly the same conditions [155]. The TV norm–based penalty or prior functions also outperform the L_2 norm-based penalty functions in terms of noise removal. In [95], the TV norm has been incorporated as it provides a fused image which is less noisy and has pronounced edges as compared to the one obtained using the L_2 norm based penalty function. The formal expression of the TV norm for image F is given by Eq. (5.13).

$$\mathscr{P}(F) \equiv TV(F) = \int_x \int_y |\nabla F| \, dx \, dy = \int_x \int_y \sqrt{F_x^2 + F_y^2} \, dx \, dy, \qquad (5.13)$$

where F_x and F_y denote the spatial derivatives of the image in the x and y directions, respectively.

We may combine the data term from Eq. (5.12), and the prior term from Eq. (5.13) into the basic formulation from Eq. (5.10) to estimate the fused image F. Some of the terms that do not affect the estimation process can be eliminated. We rewrite the Bayesian solution by appropriately introducing the regularization parameter λ_F, and eliminating the terms that do not affect the process of estimation to formulate Eq. (5.14).

$$\hat{F} = \arg\max_F \int_x \int_y \left(-||s - \beta F||^2 - \lambda_F |\nabla F| \right) \, dx \, dy. \qquad (5.14)$$

By the change of sign, we formulate the MAP estimation as the minimization problem, i.e.,

$$\hat{F} = \arg\min_F \int_x \int_y \left(||\mathbf{s} - \beta\, F||^2 + \lambda_F\, |\nabla F| \right)\, dx\, dy. \qquad (5.15)$$

Equation (5.15) represents the final solution for the estimation of the fused image. An iterative solution to Eq. (5.15) can be derived from the approach proposed by Rudin et al. [155] using the evolution parameter-

$$\hat{F}^{(m+1)} = \hat{F}^{(m)} - \tau \left(\beta^T (\beta \hat{F}^{(m)} - \mathbf{s}) + \lambda_F\, \nabla \bullet \left(\frac{\nabla \hat{F}^{(m)}}{|\nabla \hat{F}^{(m)}|} \right) \right), \qquad (5.16)$$

where (m) indicates the iteration number, and τ refers to the step size per iteration. $\bullet$ indicates the dot product operator. The iterations can be stopped when the process converges, i.e., when the difference in the fused images obtained over successive iterations ($|\hat{F}^{(m+1)} - \hat{F}^{(m)}|$) is smaller than a pre-defined threshold. The TV prior-based estimators typically converge within a very few iterations. The intensity values of estimated fused image are then linearly scaled into [0, 1] to use the complete dynamic range of the display device.

5.6 Implementation

Selection of the parameters of the algorithm is an important factor for its performance. The above discussed technique is a two-step procedure involving the computation of the sensor selectivity factor (β), and the estimation of the fused image (F). The computation of β requires evaluation of two quality measures at every pixel in the data. The first quality measure is related to the well-exposedness of the pixel which is measured through a Gaussian function. The value of variance σ_β^2 in Eq. (5.6) is essential for a proper evaluation of the measure Q_1. This variance parameter σ_β^2 defines the spread of the bell-shaped Gaussian curve. A very small value of the spread parameter results in a narrow-width Gaussian curve around the mid-intensity level. Thus, a large number of pixels that are away from the middle gray level get assigned to a very small quality values. Although we want to exclude the gray values on either extrema (by assigning them smaller possible quality values), in this scheme, one may lose the information provided by pixels that are slightly away from the mid-intensity value. On the other hand, a too high value of σ_β^2 results in a Gaussian curve with a very slow fall-off. Thus, most of the pixels, including the ones close to either extrema will also get significant weightage which is not desired. We have selected the value of this parameter to be 0.20 which gives a desirable balance. The constant C in the measure Q_2 in Eq. (5.7) is the second parameter we need to set before we proceed to actual fusion. This constant determines the selection of pixels contributing towards fusion on the basis of amount of the local sharpness. As C increases, the value of

Q_2 also increases. Thus, increase in the value of C leads to more pixels contributing towards the fusion output. One should, however, note that the value of the sensor selectivity factor β does not change in the same manner as the C or Q_2 due to explicit normalization. The exact relation between C and the quality of the fused image is dependent on the data statistic, and thus, it is difficult to obtain a simple expression relating them. One can, however, say that the value of C is related to the total number of pixels contributing towards the fusion result. We have heuristically selected C to be equal to 1.0. The regularization parameter λ_β determines the relative weightage given to the smoothness in β as compared to the product of quality measures. A too high value of λ_β brings in an excessive smoothness in β, while a very small value of the regularization parameter does not consider the effect of spatial correlation within the constituent bands. As mentioned earlier, most of the hyperspectral images (or remote sensing images) capture large areas on the earth from a high altitude. These images depict a large agglomeration of very small objects. For a better visualization, we expect these objects to be clearly identifiable in the fused image. A high value of λ_β can oversmooth small objects as smaller gradient values may get flattened. We have chosen the values of λ_β around 50 to solve Eq. (5.8).

The final stage of the Bayesian fusion in Eq. (5.16) requires the choice of two parameters- the step size τ, and the regularization parameter λ_F. These parameters govern the convergence and performance of the TV norm–based iterative procedure. The step size τ decides the magnitude of the change in the output per iteration. A process with very small values of the step size turn out to be quite slow as they require a large number of iterations to converge. High values of the step size may lead to instability in the iterative process. We have chosen the step size τ to be small, around 0.20 for the stability considerations. The choice of regularization parameter λ_F is critical as it decides the relative weightage of the total variation (TV) term, i.e., the prior, with respect to the weightage associated with the image formation model, i.e., the data fitting term. High values of λ_F in Eq. (5.15) indicate a much higher weightage of the TV term as compared the image model. This excessive domination may result in flattening of the output by removal of the small textures/features. We have chosen this value to be in the range of 10^{-2} to 10^{-3} which provides enough weightage to the TV term to take into account the spatial correlation within the fused image. One of the typically employed procedure to stop the iterations is based on the change in the successive values of the minimization functional. We have employed a similar stopping criteria in this work, that is comparing the relative difference in the minimization functional per iteration. Results were typically obtained after 6–8 iterations as the process converged.

5.7 Experimental Results

In this monograph we have already explained a different solution for the problem of fusion of hyperspectral images. In the next two chapter we present the readers with two more methods for the hyperspectral image fusion. We believe that putting up the results from all the fusion techniques together facilitates a better comparative analysis to the readers both in a subjective and objective terms. Therefore, we shall present combined results of fusion over several hyperspectral datasets using these techniques explained in this monograph along with some other recently developed techniques in a later chapter. However, in this chapter we provide the results of fusion using the Bayesian technique over a single dataset for a quick illustration.

Consider the urban hyperspectral data used in the previous chapter again. These data depict some region of Palo Alto, CA that has several urban establishments. The data contain 242 bands of dimensions (512×256) each. We initially remove the noisy and zero-response bands as done for the previous solution. We process the entire dataset to obtain a grayscale fused image. First we compute the sensor selectivity factor β for every pixel in the data which is the normalized product of two quality measures. The second step consists of an iterative procedure to generate the fused image F. The fused image so obtained represents the best possible image estimated under the given constraints for the given hyperspectral data. This resultant image, as shown in Fig. 5.2a is a combined response of the scene over an approximate band-width of 2200 nm. The fusion results are also often provided as color (RGB) images as they provide a better visual interpretation of the scene. Most of the fusion solutions for hyperspectral data generate the color images using some pseudo-coloring schemes. In order to illustrate the RGB versions, we first partition the input data into three subsets. While there could be various strategies for this partitioning, we follow a simple one. We obtain three groups by sequentially partitioning the data along the wavelength such that every groups contains nearly a similar number of bands. These three groups are processed independently using the Bayesian technique to generate the corresponding three fused images. These images are then assigned to the red,

Fig. 5.2 Results of Bayesian fusion of the urban image from the Hyperion. **a** Grayscale fused image, and **b** RGB fused image (©2013 Elsevier, Ref. [91]). (Color is viewable in e-book only.)

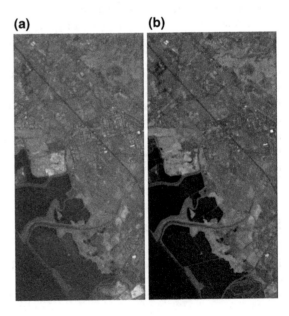

(a)   **(b)**

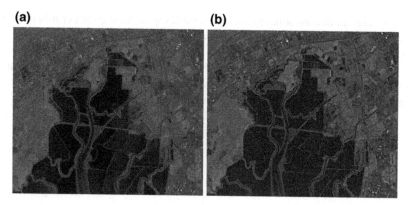

Fig. 5.3 Results of Bayesian fusion of the moffett$_2$ image from the AVIRIS. **a** Grayscale fused image, and **b** RGB fused image (©2013 Elsevier, Ref. [91])

green, and blue channels of the display or printing device to generate the RGB composite output. Figure 5.2b shows the RGB version of the fusion result for the urban data. It should, however, be noted that the colors in this image have been assigned by a pseudo-coloring scheme as described, and they may not have any resemblance to their natural meaning.

We also provide the results of Bayesian fusion for the moffett$_2$ data from the AVIRIS. These data depict a very large number of small objects such as urban establishments, and hence, we are particularly interested in observing how these objects are represented in the resultant fused image. Figure 5.3a shows the grayscale version of the result of the Bayesian fusion. The RGB version of the fusion output is shown in Fig. 5.3b where each of the R, G, and B bands are generated from fusion of nearly 1/3-rd bands of the moffett$_2$ hyperspectral image.

5.8 Summary

This chapter presents a Bayesian solution for fusion of hyperspectral images where the primary purpose of fusion is quick and efficient visualization of the scene contents. The fusion methodology relies on the statistical model of image formation which relates observed hyperspectral bands to the fused image through a first order approximation. This model uses a parameter called the sensor selectivity factor (β). We have discussed an approach for the computation of these parameters based on some of the characteristics of the hyperspectral data. A Bayesian framework has been discussed for the estimation of the fused image using the MAP estimator where the TV norm-based prior has been employed. The fusion technique being completely data-driven, does not require any information related to the sensor device, and hence, it tends to be generic and sensor or device independent. The use of Bayesian framework not only provides an elegant solution, but also enables us to handle the perturbations in the corresponding model in a very efficient manner.

The presented technique provides a computationally simple and iterative solution. The main focus lies in providing a statistical framework for hyperspectral image fusion. A variety of different quality factors, and smoothness norms can be employed within this framework without changing the solution modality.

Chapter 6
Variational Solution

6.1 Introduction

We have studied pixel-based image fusion as a linear combination of multiple input images. The weights, or more precisely fusion weights, are the data-dependent terms as they have been calculated from the set of input images (or input hyperspectral bands in our case). For example, the bilateral filtering–based fusion technique calculates the fusion weights (w) using a predefined function. The Bayesian fusion technique is based on the computation of the sensor selectivity factor (β) which indicates the contribution of each pixel toward the fused image. Both of these fusion techniques *explicitly* calculate the weights as a function of the input hyperspectral data. These functions are also referred to as the weighting functions, while the weights are more commonly known as the α matte in the graphics literature. The fusion weights act as intermediate variables of the fusion process that define the relationship between the fused image and the input hyperspectral bands. The purpose of the weighting function which generates the fusion weights, is to implicitly specify the model for the fusion process. An explicit computation of fusion weights is, therefore, not required so long as the underlying model is well specified. In other words, we do not necessarily have to compute the fusion weights independently, if we can appropriately model the weighting function as a data-dependent term to weigh the hyperspectral bands. We now explore this possibility, and show how a fusion technique can be developed without any explicit calculation of fusion mattes. In Chap. 5, we have seen that imposing some constraints through a prior related to the smoothness of the output fused image, gives it a natural and visually pleasing appearance. In order to incorporate the smoothness constraint, we adopt an approach based on calculus of variations. This chapter discusses how we can start with an initial estimate of the fused image, and iteratively converge to obtain the desired resultant image based on certain constraints as well as the predefined weighting function, without ever explicitly computing the weights.

In case some of the readers are not familiar, we discuss in very brief about the calculus of variations in Sect. 6.2, which will be used as a tool to address the fusion

S. Chaudhuri and K. Kotwal, *Hyperspectral Image Fusion*,
DOI: 10.1007/978-1-4614-7470-8_6, © Springer Science+Business Media New York 2013

problem. This section also describes the Euler-Lagrange equations which are important tools for solving the problems based on the calculus of variations. The design of the weighting function, and the formulation of the problem in a variational framework are provided in Sect. 6.3. Sections 6.4 and 6.5 provide the implementation details, and the corresponding results of fusion over a couple of hyperspectral datasets. Summary is presented in Sect. 6.6.

6.2 Calculus of Variations

Computer vision usually deals with problems that are ill-posed in Hadamard sense. Hadamard defined a mathematical problem to be well-posed when the solution [139] *(i)* exists, *(ii)* is unique, and *(iii)* depends continuously on the input data. The problems that are not well-posed in Hadamard sense, are said to be ill-posed. Generally, inverse problems are ill-posed. A few examples of ill-posed problems are estimation of structure underlying the scene, computer tomography, and super-resolution. The ill-posed problems are often converted into well-posed problems by introducing appropriate constraints. This process forces the solution to lie in that subspace where it has been well defined [139]. This process is known as the regularization which reformulates the problem for a better numerical analysis. Consider an ill-posed problem of finding ζ from the data ρ where they are related by the model A, such that $A\zeta = \rho$ (The ill-posedness of the problem could be due to either the non-existence or ill-conditioning of A^{-1} when ζ may not exist). The most commonly used regularization technique in such cases is to select a suitable norm, and introduce an additional constraint over ζ in the form of a stabilizing functional Γ. This process is known as the Tikhonov regularization. The regularization term is often weighted in order to decide the preference between the particular constraint and the data fitting term. Thus, the overall problem can be formulated as the minimization of the following functional:

$$||A\zeta - \rho||^2 + \lambda\,||\Gamma\zeta||^2,$$

where the scalar λ controls the relative weightage given to the constraint against the data fitting term. It is known as the regularization parameter. The choice of regularization functional is guided by a physical analysis of the system, and its mathematical consideration.

Since the early days of computer vision, smoothness is one of the most commonly employed constraints. The natural surfaces are generally smooth and so are the natural images except at edges. This choice is popular because of its- *(i)* realistic and practical nature, *(ii)* physical interpretation, and *(iii)* mathematical elegance. The last point refers to the fact that the smoothness can be easily incorporated by penalizing large deviations, i.e., by penalizing the higher values of the derivative. Smoothness constraint has been adopted in several algorithms, few examples of which include shape from shading [76, 204], stereo analysis, optical flow estimation [8, 125], image compositing [108, 145], active contours [85], and super-resolution [82].

These formulations are composed of an appropriate data fitting term and the regularization term to incorporate the smoothness constraint.

The formulation of the aforementioned set of problems involves a functional or a function of functions, and their derivatives, to be integrated over a finite image domain. The objective is to determine an appropriate unknown function that maximizes or minimizes the corresponding functional. A functional Ψ over a function ω in variable ζ is defined by Eq. (6.1).

$$\Psi(\omega) \equiv \int_{\zeta_1}^{\zeta_2} \psi\left(\omega(\zeta), \omega'(\zeta), \omega''(\zeta), \cdots\right) d\zeta, \qquad (6.1)$$

where $\omega'(\zeta)$ and $\omega''(\zeta)$ represent the first and second derivatives of the function ω with respect to ζ, respectively, and ζ_1 and ζ_2 define the limits of integration. Let the function ω be defined within some space of functions $\mathbb{F}$, i.e., $\omega \in \mathbb{F}$, and $\zeta \in \mathbb{R}$, then the functional Ψ defines the mapping $\Psi : \mathbb{F} \to \mathbb{R}$.

The problem of finding stationary functions for which $\Psi(\omega)$ is the minimum, is of interest in several fields of computer vision, optimization, control theory, and physics. This task can be accomplished by solving the Euler-Lagrange equation associated with the corresponding functional.

Consider the functional Ψ defined in Eq. (6.1). One would like to find the unknown function ω, defined over $[\zeta_1, \zeta_2] \subset \mathbb{R}$. The Euler-Lagrange equation corresponding to this functional is given by Eq. (6.2).

$$\frac{\partial \psi}{\partial \omega} - \frac{\partial}{\partial \zeta}\left(\frac{\partial \psi}{\partial \omega'}\right) + \frac{\partial^2}{\partial \zeta^2}\left(\frac{\partial \psi}{\partial \omega''}\right) - \cdots = 0. \qquad (6.2)$$

If the functional in Eq. (6.1) involves only first derivative, the corresponding Euler-Lagrange equation is given by Eq. (6.3).

$$\frac{\partial \psi}{\partial \omega} - \frac{\partial}{\partial \zeta}\left(\frac{\partial \psi}{\partial \omega'}\right) = 0. \qquad (6.3)$$

The regularization parameter, if any, is also incorporated into this equation, as it will be seen in the next section.

The Euler-Lagrange equation can be extended to the functionals of multiple functions, as well as multiple variables. In 2-D, the cost functional $\Psi(\omega)$ takes the form

$$\Psi(\omega) \equiv \iint \psi(\omega(x, y), \omega_x(x, y), \omega_y(x, y)) \, dx dy, \qquad (6.4)$$

where $\omega_x(x, y)$ and $\omega_y(x, y)$ are derivatives of ω along the x- and y-directions, respectively. The corresponding Euler-Lagrange equation is given by Eq. (6.5).

$$\frac{\partial \psi}{\partial \omega} - \frac{\partial}{\partial x}\left(\frac{\partial \psi}{\partial \omega_x}\right) - \frac{\partial}{\partial y}\left(\frac{\partial \psi}{\partial \omega_y}\right) = 0. \qquad (6.5)$$

For image processing and vision problems, we mostly deal with functionals and Euler-Lagrange equations of two variables. The usefulness of these equations will be evident in the next section where these have been employed to obtain the resultant fused image.

6.3 Variational Solution

We would like to develop an algorithm that will iteratively produce the resultant image that preserves most of the salient features of the set of input hyperspectral bands. We want to design the fusion weights $w_k(x, y)$ which are essentially the α-mattes for every band in the data such that the said goal is achieved. The pixels with a high local contrast provide a good amount of visual information to the observer. Also, features such as edges and boundaries possess the characteristic of high local contrast due to significant differences in gray values from their local neighborhood. Therefore, the choice of local contrast as the weighing parameter is quite justified. The contrast being a dissimilarity measure, these pixels might be quite dissimilar in gray values. However, we want the merging process to be smooth, providing a natural appearance to the resultant fused image.

One of the easy ways to generate the α-matte is to select just one pixel from all available bands at every spatial location using certain quality criterion, and the contribution from the rest of the bands is zero. The fusion weight $w_k(x, y)$ takes one of the values in $\{0, 1\}$ at any location (x, y) in any hyperspectral band I_k. This approach has two major shortcomings. First, this approach essentially selects only one pixel per spectral array, and therefore, a large amount of data is discarded. In order to make the solution efficient, one has to select the quality measure carefully. Often, a set of multiple quality measures is needed as a single measure may not be able to efficiently capture the saliency of the pixel data. Secondly, when the choice of matte selection depends on certain quality criterion of the fused image, the fusion procedure leads to a combinatorial search problem which is computationally demanding. We can, however, as explained in previous two chapters, relax the condition on fusion weights $w_k(x, y)$ allowing it to take real values in the range $[0, 1]$. The fusion weights have to satisfy the usual properties, i.e., non-negativity, and unity sum. As the fusion weights are restricted to the aforementioned range, these are essentially non-negative by formulation. The normalization condition refers to the summation of all weights along the spectral dimension for every spatial location (x, y) being equal to one. We have to impose the normalization constraint explicitly, i.e.,

$$\sum_{k=1}^{K} w_k(x, y) = 1 \quad \forall(x, y).$$

Subject to this constraint, we can employ the variational framework to generate a fused image that combines input hyperspectral bands where fusion weights have been derived from certain measure related to the local contrast. A similar approach has been employed for fusion of multi-exposure images to form a high dynamic range (HDR)-like image in [145]. The basic formulation of the fusion problem using a variational framework is provided in Eq. (6.6).

$$
F(x, y) = \arg\min_{F} \int_x \int_y \left\{ \left(F(x, y) - \sum_{k=1}^{K} w_k(x, y) \, I_k(x, y) \right)^2 \right.
$$
$$
\left. + \lambda_{\mathrm{var}} \left(\frac{\partial F(x, y)^2}{\partial x} + \frac{\partial F(x, y)^2}{\partial y} \right) \right\} dx\, dy, \qquad (6.6)
$$

where λ_{var} is the regularization parameter, and F is the fused image. This parameter is set by the user to obtain a desired balance between the weightage assigned to the matte (first term in Eq. (6.6)), and the weightage assigned to the smoothness of the fused image. The second term in the above Equation is known as the regularization term. In the present case of hyperspectral image fusion, we want the input bands to merge smoothly without producing any visible artifacts in the fused image. As most natural images have a moderate to high amount of spatial correlation, any sharp and arbitrarily discontinuous region generated via fusion may turn out to be a visible artifact. In order to avoid such artifacts, we would like the fused image to have a certain degree of smoothness. One of the ways of accomplishing this objective is to penalize the departure of the fused image from smoothness which can be characterized by high values of its spatial gradient. We incorporate this in the form of a regularizing term, and seek for the minimization of the corresponding functional defined in Eq. (6.3). It may, however, be noted that several other types of regularization or penalty functions have been proposed in the literature. The choice of the regularization function is determined by nature of the data, and is of utmost importance for obtaining the desired output.

We have the basic formulation of our fusion problem as given by Eq. (6.6). Now, we shall focus on designing an appropriate function for the calculation of fusion weights. Pixels with higher local contrast bring out a high amount of visual information. We quantify the local contrast of a given pixel by evaluating the variance in its spatial neighborhood. Thus, fusion weights $w_k(x, y)$ for the k-th hyperspectral band should be directly proportional to the local variance $\sigma_k^2(x, y)$. However, in case of homogeneous regions in hyperspectral bands, the variance becomes zero, and such regions may not get an appropriate representation in the final fused image. To alleviate this problem, we incorporate a positive valued additive constant C' to the weighting function. The constant C' serves the similar purposes as the constant C does in the case of bilateral filtering–based fusion described in Chap. 3. Apart from the contrast, we also want to balance the saturation level in the resultant fused image. The fused image should have very little under- or over-saturated regions. It is also desired that the fused image be quite close to the mean observation along the band for radiometric consideration. Hence, we consider the weight $w_k(x, y)$ to be proportional to $(F(x, y) - I_k(x, y))^2$, with the proportionality constant being equal to the measure of local contrast. A similar weighting function based on the intensity difference has been proposed in [102, 145]. It is also interesting to note that if K hyperspectral bands are considered to be different noisy (assumed Gaussian) observations as discussed in the previous chapter, then this solution would lead to an optimal noise smoothing filter. The fusion weights can now be computed using Eq. (6.7).

$$w_k(x, y) = \left((\sigma_k^2(x, y) + C')(F(x, y) - I_k(x, y))^2\right). \tag{6.7}$$

It may be noted that one may employ various other quality measures such as spatial gradient to evaluate the saliency of the pixel. These fusion weights, by definition, are guaranteed to be non-negative. However, we should explicitly normalize them for every pixel. The normalized fusion weights are given by Eq. (6.8) as,

$$w_k(x, y) = \frac{\left((\sigma_k^2(x, y) + C')(F(x, y) - I_k(x, y))^2\right)}{\sum_{k=1}^{K}\left((\sigma_k^2(x, y) + C')(F(x, y) - I_k(x, y))^2\right)}, \tag{6.8}$$

which is merely dividing by the sum of all weights at a given pixel over all input images to ensure that the weights sum to unity for a given pixel (x, y).

We have transformed the fusion problem into an optimization problem where we seek to obtain the resultant image that iteratively refines itself in order to merge input hyperspectral bands smoothly. We provide the solution to this problem of calculus of variations using the corresponding Euler-Lagrange equation described in the previous section. The basic variational formulation is given in Eq. (6.6), while the fusion weights $w_k(x, y)$ are computed using Eq. (6.8). For the purpose of brevity, we omit the arguments x and y, and restate the same for a quick reference in Eq. (6.9).

$$F = \arg\min_F \int_x \int_y \left\{ \left(F - \sum_{k=1}^{K} w_k I_k\right)^2 + \lambda_{\text{var}}\left(||F_x||^2 + ||F_y||^2\right) \right\} dx\, dy, \tag{6.9}$$

where F_x and F_y refer to the partial derivatives of the fused image F with respect to x and y, respectively. At every iteration, the solution generates the output image F which progressively minimizes the above expression. The newly formed fused image is subjected to further refinement during the next iteration till a satisfactory result is obtained. The discretized version of the solution of the Euler-Lagrange equation is given by Eq. (6.10).

$$F^{(m+1)} = \bar{F}^{(m)} - \frac{1}{\lambda_{\text{var}}}\left(\left(F^{(m)} - \sum_{k=1}^{K} w_k I_k\right)\left(1 - \sum_{k=1}^{K} I_k \frac{A^{(m)} D_k^{(m)} - E^{(m)} B_k^{(m)}}{A^{(m)2}}\right)\right) \tag{6.10}$$

where

$$D_k^{(m)} \equiv D_k^{(m)}(x, y) = 2 \left((\sigma_k^2(x, y) + C') \, (F^{(m)}(x, y) - I_k(x, y)) \right), \quad (6.11)$$

$$B_k^{(m)} \equiv B_k^{(m)}(x, y) = (\sigma_k^2(x, y) + C') \, (F^{(m)}(x, y) - I_k(x, y))^2, \quad (6.12)$$

$$A^{(m)} \equiv A^{(m)}(x, y) = \sum_{k=1}^{K} B_k^{(m)}, \quad (6.13)$$

$$E^{(m)} \equiv E^{(m)}(x, y) = \sum_{k=1}^{K} D_k^{(m)}, \quad (6.14)$$

where (m) indicates the iteration index, and $\bar{F}$ denotes the average value of F over its nearest 4-connected neighborhood. The variables A, B, D, and E are computed at each iteration. It may be noted that B is same as the un-normalized fusion weights, and A refers to the summation of the fusion weights for the process of normalization. As the resultant image F changes at every iteration, these variables also need recomputation at the end of every iteration. This slows down the whole fusion process.

6.4 Implementation

The variational technique described in this chapter is conceptually very simple. Additionally, it requires only two input parameters from the user- the constant C' (used in conjunction with variance $\sigma_k^2(x, y)$ to define fusion weights), and the regularization parameter, λ_{var}. We have heuristically chosen the value of C' to be 50 which was found to provide good visual results by appropriately balancing the weightage of the local variance at the pixel. As most of the natural images are smooth, one may assume the final fused image also to be smooth. In case of noisy hyperspectral bands, a higher value of λ_{var} can be used to generate a smoother final resultant image. For the illustrations in this monograph, we have assigned the value of λ_{var} to be of the order of 1–10. Higher values of λ_{var} tend to smooth out fine textural details which are important to preserve the overall contrast. For most test datasets, the algorithm was found to converge within 10–12 iterations, when the stopping criteria employed was to observe the relative change in the cost function over successive iterations.

6.5 Experimental Results

Like in the previous two chapters, we provide a couple of illustrative results of fusion using the variational technique. We have provided the results over the same hyperspectral datasets used in the previous chapter in order to maintain uniformity. More results along with their quantitative analysis are provided in Chap. 10.

Fig. 6.1 Results of fusion for
the variational approach for
the urban image data from the
Hyperion. **a** Grayscale fused
image, and **b** RGB fused
image

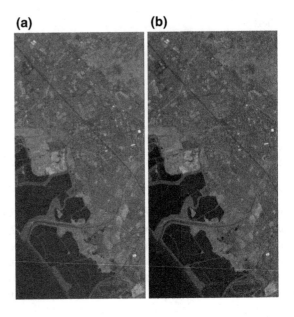

(a)   **(b)**

First we consider the urban hyperspectral image which depicts some urban regions
of Palo Alto, CA in the form of contiguous 242 bands. The dimension of each band
is (512×256) pixels. For fusing the entire hyperspectral image into a grayscale
output, the entire set of input bands is processed in a single step. The local variance
of the data needs to be computed only once, which remains constant throughout
the fusion process. The weight requires re-computation at every iteration as it is
dependent on the newly formed (intermediate) fused image. Figure 6.1a depicts the
result of fusion using the variational solution. This image represents fusion over the
spectral bandwidth of nearly 2200 nm. The procedure to generate the RGB fused
image remains same as the one employed for the other two methods presented in the
earlier chapters. We partition the input hyperspectral data into 3 subsets of contiguous
bands, each nearly of the same cardinality, i.e., each subset contains nearly $1/3$-rd of
the hyperspectral data. These three subsets are then fused independently using the
variational technique to yield three corresponding fused images. The RGB composite
image is formed by assigning these fused images to the red, green, and blue channels
of the standard display. The corresponding RGB image is shown in Fig. 6.1b. Both
images in Fig. 6.1 appear visually smooth, and do not produce any visible artifacts.
However, at certain places, the edges and boundaries of various objects in the data
appear less sharper. These features represent the high frequency components in the
image which get attenuated as the regularization term [Eq. (6.9)] penalizes sharp
changes or high frequency components in the output.

   We have used the moffett$_2$ hyperspectral image from the AVIRIS as the second test
data. The grayscale output image obtained by the variational method for the moffett$_2$
data is shown in Fig. 6.2a. This image is the result of combining 224 hyperspectral

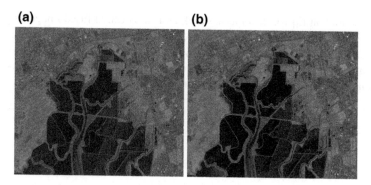

Fig. 6.2 Results of fusion for the variational approach for the moffett$_2$ image from the AVIRIS. **a** Grayscale fused image, and **b** RGB fused image

bands each of which has a nominal bandwidth of 10 nm. The RGB version of the fusion output is obtained by independently fusing three nearly equal subsets of the input hyperspectral data, and then assigning it to the different channels of the display. As previously stated, we have followed a simple strategy for partitioning the set of hyperspectral bands, that is, partition the sequentially arranged bands into nearly equal three subsets. The RGB version of the result for the moffett$_2$ data is shown in Fig. 6.2b. One may notice that the fusion results are quite good, although some loss of details may be seen at small bright objects due to smoothness constraint imposed on the fused image. For improved results, one may use a discontinuity-preserving smoothness criterion as suggested by Geman and Geman [63]. However, this would make the solution computationally very demanding and hence, is not pursued.

6.6 Summary

A technique for fusion of hyperspectral images has been discussed in this chapter, without *explicit* computation of fusion weights. The weighting function, defined as a data dependent term, is an implicit part of the fusion process. The weighting function is derived from the local contrast of the input hyperspectral bands, and is also based on the concept of balancing the radiometric information in the fused image. An iterative solution based on the Euler-Lagrange equation refines the fused image that combines the pixels with high local contrast for a smooth and visually pleasing output. However, at some places, the output appears oversmooth which smears the edges and boundaries.

The solution converges to the desired fused image within a few iterations. However, it requires computation of the implicit weighting function at the end of every iteration which makes it computationally expensive as compared to the techniques described in the previous chapters.

A close look at Eq. (6.9), solution of which is the desired fused image, suggests that the fused image F is no longer a convex combination of all input hyperspectral bands, i.e., $F(x, y) \neq \sum_k w_k(x, y) I_k(x, y)$. Hence the weights $w_k(x, y)$ no longer define an α-matte, unlike other techniques discussed in this monograph. Hence, this technique may be called a *matte-less* formulation of fusion problem.

Chapter 7
Optimization-Based Fusion

7.1 Introduction

We have explored three hyperspectral image fusion techniques in the earlier chapters of the book. We have also reviewed several other hyperspectral image fusion techniques in Chap. 2 along with a brief overview of a number of generalized image fusion schemes. The common feature of most of the existing fusion methodologies is that the fusion rule operates over spatial characteristics of the input images (or hyperspectral bands) to define the fusion weights. Fusion weights which have been calculated from some kind of saliency measure of the pixel, determine the effective contribution of the corresponding pixel toward the final result. Thus, the characteristics of the input data control the process of fusion, and hence, the resultant-fused image. As mentioned earlier, in this monograph, we are focused on obtaining the fused image for the purpose of visualization of the scene contents. In such a case, would it not be highly desirable that the fused image should possess certain characteristics to facilitate a better visualization and interpretation of the scene? In other words, it is beneficial to have a fusion scheme that takes into consideration the characteristics of the output, rather than focusing on the input characteristics as was done in previous chapters. Although we have considered smoothness of the fused image, we have not given specific attention to any other qualities the fused image should possess. In this chapter, we explain some of the desired elements of the image quality. Then we explain the formulation of a multi-objective cost function based on some of these elements. This cost function has been developed into the variational framework which has been explained in the previous chapter (Sect. 6.2, Chap. 6). Finally, we discuss how an iterative solution can be formed using the Euler-Lagrange equation.

The next section discusses various aspects of image quality, and their usage in image enhancement. The multi-objective cost function, and its solution are provided in Sect. 7.3. The corresponding details of implementation are given in Sect. 7.4. Results of fusion are illustrated in Sect. 7.5. Finally, Sect. 7.6 brings out the summary.

S. Chaudhuri and K. Kotwal, *Hyperspectral Image Fusion*,
DOI: 10.1007/978-1-4614-7470-8_7, © Springer Science+Business Media New York 2013

7.2 Image Quality

Various aspects of image quality are useful to objectively quantify how good or well-perceived the image is. These aspects can be different depending on the context and requirements of the application. Let us look at some of the common aspects of an image that are attributed to its quality.

1. **Spatial Resolution**: Spatial resolution of an image refers to the size of the smallest object that can be detected or resolved. For digital images, the size of the smallest resolvable object cannot be smaller than the pixel size. Therefore, the spatial resolution is limited by the pixel size. The resolution is mainly determined by the instantaneous field of view (FOV) of the imaging sensor which quantifies the area of the scene viewed by the single sensor element. Images are said to have fine or high resolution when each of the sensor elements *sees* a small area of the scene, and thus, smaller objects can be easily discriminated. In coarse or low resolution images, only large features are clearly visible. However, the terms high or low, and fine or coarse are relative to the subject and the context. They do not have unified definitions. In the case of remote sensing images, the spatial resolution refers to the smallest area on the earth's surface being *seen*, or captured by the single sensor element. The altitude of the imaging device, i.e., its distance from the target area is an important factor in determining the spatial resolution which differentiates remote sensing images from other ones. Since the sensors are typically located at a high altitude, they can capture a large geographical area, but cannot provide minor spatial details.

   Images with higher spatial resolution are always desirable since they have a better visual appearance for human observers. Also, higher information content is useful for processing of machine vision and pattern classification algorithms. The technique of enhancing the spatial resolution of the image beyond what is provided by the imaging sensor with simultaneous reduction or elimination of aliasing or blurring is known as super-resolution [30]. Most of the super-resolution techniques combine the information from multiple low resolution observations of the scene to generate a single high resolution image, while single-frame techniques process only a single low-resolution observation for the purpose of super-resolution. Various approaches to single-frame and multi-frame super-resolution can be found in [17, 30, 31, 123, 127].

   In the case of remote sensing images, the need for increasing spatial resolution is often felt in the case of multispectral images. This is achieved by integrating information from a panchromatic image having a higher spatial resolution. This process, known as pan-sharpening has been discussed in Sect. 2.1.2. However, the pan-sharpening process is not yet found to be much relevant for hyperspectral images.

2. **Gray-Level Resolution**: The radiometric or gray-level resolution is defined as the smallest discriminable change in the gray level [65]. It is dependent on the signal to noise ratio (SNR) of the imaging device. However, in digital images, this factor is limited by the number of discrete quantization levels used to digitize

the input signal. This quantity is related to the number of gray levels in the image. An image quantized using 8-bits can accommodate 256 distinct intensity values, i.e., gray levels. The quantization using a fewer number of bits degrades the radiometric resolution, loses the accuracy, and hence the quality of an image. Images quantized with a higher number of bits appear vivid as they cover a higher dynamic range. While most commonly used images use 8-bit quantization, the sensors that can capture 10-bit, 12-bit, and 16-bit have been manufactured. Several hyperspectral sensors provide the reflectance response in the form of a 12-bit data. For a better exploitation of this high dynamic range images, however, one requires a device with a similar displaying capability. The real world scenes are sometimes captured in the form of multiple observations obtained by varying the parameters of the imaging system in order to cover as large a dynamic range as possible. These multi-exposure observations are then blended together to provide a *feel* of high dynamic range (HDR) image to the viewer. The technique of compositing an HDR-like image from multiple observations captured from a low dynamic range (LDR) device by varying the exposure time is an interesting research area in computational photography. Various methodologies for compositing include blending regions from different LDR images [67], variational formulation [145], maximum likelihood (ML) estimation [108], and exposure fusion [113]. Since most of the hyperspectral data already comes with a fine quantization level, this aspect is not important for hyperspectral image visualization.

3. **Sharpness**: While image sharpness can be interpreted in different ways, it is a measure of clarity of detail in an image. High frequencies in the image correspond to the detail, and hence some definitions of sharpness are related to the modulation transfer function (MTF) of the imaging device which represents the magnitude of the normalized spatial frequency response [80]. The perceived sharpness of the image is related to the spatial frequency as a function of MTF. Sometimes the perceived sharpness is associated with the quality of edges and boundaries which have significant high frequency components. This perceived sharpness is affected by factors such as spatial resolution and acutance [98]. As the spatial resolution describes the ability of discrimination of finer objects, images with higher spatial resolution can provide a better delineation of an object due to reduced aliasing. Acutance describes how quickly image information changes at an edge, and thus, a high acutance results in sharp transitions and detail with well-defined boundaries. Acutance can be calculated from the mean square density gradient across the edge [3].

Sharpening enhances the perceived details by creating more pronounced edges and boundaries in the image. The unsharp masking has been one of the most popular techniques of image sharpening, especially used in printing industry. The low pass filtered version of an image is compared with the original for selective enhancement of the detail components [110]. The performance of unsharp masking can be improved using adaptive filters [141], or non-linear filters [148]. Software packages such as Adobe® Photoshop® and GIMP also provide sharpening facilities. However, over-sharpening produces visible edge artifacts, known as halos, and makes an image appear granular.

4. **Contrast**: It is another attribute which makes an object distinguishable from the rest of the image contents. Contrast is largely related to the image sharpness. Often, the perceived sharpness refers to the mixture of spatial resolution, acutance, and contrast [98]. As the contrast is ultimately perceived by the human observers, it is often defined based on the human visual system (HVS). Several definitions of contrast based on HVS co-exist in the literature. The Weber contrast is defined as the ratio of the difference between luminance of the object and that of the background to the background luminance [80]. The Michelson contrast is defined as the ratio of half value of the available luminance range to the sum of this half value and the object luminance [128]. The dynamic range of this measure is [0, 1], as opposed to the former case of Weber contrast where it is [−1, ∞]. Peli [128] discussed the RMS contrast which does not involve the HVS factors. The RMS contrast is defined as the standard deviation of the image intensity. Variance of an image is also used as one of the contrast indicators due to ease of computation, and non dependency on any other factors. Lee has proposed contrast enhancement of images based on their local mean and variance [99]. The variance has been used as the performance measure for several image enhancement techniques including image fusion.

5. **Exposedness**: It refers to the clipping of intensity levels due to finite length storage space per pixel. The real world scene has a large dynamic range. In order to compress it within the limits dictated by the storage and/or display system, the intensity (gray) values above the pre-decided maximum, and the values below the pre-decided minimum have to be clipped to the respective gray levels. This operation gives rise to a large number of pixels with improper intensity values, and reduces the information content in the image. An incorrect exposure time of the imaging device gives rise to over- or under-saturated regions. This phenomenon is also referred to as over- or under-exposure in photography. Too bright or too dull images are not visually appealing. It can also be easily understood that saturated regions lack contrast, and hence it reduces image clarity.

   The maximum of Laplacian has been considered as the measure of saturation in [16] for fusion of monochrome and color images. In [113], well-exposedness of the scene has been regarded as one of requirements while obtaining visually appealing images, which is measured as the closeness of the pixel to the mean gray value.

We shall be using some of these image quality measures while designing the fusion strategy. Section 7.3 explains the formulation of a multi-objective function based on some of the characteristics discussed in this section. The solution of this optimization problem has been provided using the Euler-Lagrange equation which are among most popular tools for solving problems in variational framework. We have already provided a brief overview of the calculus of variations and the Euler-Lagrange equation in Sect. 6.2. We have also demonstrated their use in developing a fusion solution that does not require explicit computation of fusion weights. In this chapter, we employ the variational framework to develop a multi-objective cost function based on the properties of the fused image.

7.3 Optimization-Based Solution to Fusion

We want to obtain a single image representation of the hyperspectral data cube by selectively merging the useful features from the set of source images. When the final image is formed as a linear combination of the pixels from all of the input image bands, which is the usual case, no properties of the resultant fused image are being explicitly considered in most of the existing fusion methodologies. As our primary objective is the visualization-oriented fusion of the hyperspectral images, it is highly desirable to use a fusion strategy which generates the resultant image with a certain characteristics. The fusion technique presented here is based on some of the aforementioned characteristics of the fused image. Based on these, we discuss how a multi-objective cost function has been developed in order to transform the fusion problem into an optimization framework in the next subsection. The solution provides an optimal set of weights for the purpose of fusion of the image bands as has been explained subsequently.

7.3.1 Formulation of Objective Function

The basic approach toward the generation of the fused image remains the same as that of any pixel-based technique where the fused image is a linear combination of the input spectral bands. The different sets of weights generated by different fusion techniques are also popularly known as α-mattes in computer graphics literature. We have already discussed the primary expression of fusion using these α-mattes in earlier chapters of the book.

The fused image $F(x, y)$ of dimension $(X \times Y)$ is generated from a linear combination of the observations across spectral bands at every pixel location as represented by Eq. (7.1).

$$F(x, y) = \sum_{k=1}^{K} \alpha_k(x, y) I_k(x, y) \quad \forall (x, y), \tag{7.1}$$

where $\alpha_k(x, y)$ is the value of α-matte for the corresponding pixel which acts as the fusion weight. The fusion weights $\alpha_k(x, y)$ should satisfy the following properties-

1. At any given pixel, the sum of all the weights should be equal to unity, i.e.,

$$\sum_{k=1}^{K} \alpha_k(x, y) = 1 \quad \forall (x, y).$$

2. The weights should be non-negative, i.e.,

$$\alpha_k(x, y) \geq 0 \quad \forall (x, y).$$

The non-negativity of the weights provides a sufficient condition for the fused image $F(x, y)$ to be also non-negative.

Throughout the fusion procedure, the input hyperspectral data is assumed to be normalized such that, $0 \leq I_k(x, y) \leq 1 \, \forall \, (x, y)$. In the present work, we are dealing with the visualization-oriented fusion of hyperspectral images, where a human analyst will observe the contents of the scene. While almost all of the elements of image quality are desirable in a single image—this methodology does not focus on the enhancement of spatial or radiometric resolution. The focus is on other elements that one would like a fused image to possess. As high values of contrast make visualization appealing and clear, it is one of the desired properties of the fused image. This is helpful for some of the post-fusion operations such as object detection and segmentation apart from an enhanced visualization. A high contrast can be obtained from the pixels that are far apart in terms of their gray values. However, this may lead to a high amount of saturation (over- or under-exposedness) creating too bright or too dull regions in the fused image. This leads to loss of information, and deteriorates the quality.

The fused image is to be obtained from the hyperspectral bands which collectively encompass a large bandwidth. The reflectance response of the scene, which is a function of wavelength, also varies significantly over the set of captured bands. Through fusion algorithm, we want to map pixels from this large dynamic (intensity) range into a smaller range of the display system. This mapping can be accomplished using two different strategies:

- **S-I**: Most commercial displays are capable of providing 256 gray levels (which can be normalized into a range $[0, 1]$). The images with higher dynamic range may extend well beyond this range. If the pixel intensities beyond the maximum possible display gray level are clipped in order to accommodate into the available 8-bit display, several regions in the scene appear over-exposed or over saturated. The information within these regions gets lost. On the other extreme, when the pixel values lesser than the minimum display intensity are clipped, the scene depicts many under-exposed areas. These appear dark and dull, and thus, are not much useful for observation or any further processing. We can map the pixels from the input toward the central region (0.50) of the dynamic range of the gray values around $[0, 1]$. This mapping helps in bringing the under and over exposed regions in the scene into the mid-range of the display. The minimization of the distance of gray values from the middle of dynamic range has, in fact, been used as one of the objectives for an enhanced visualization of the image [137]. With this mapping, the dynamic range of the fused image directly gets mapped to the mid-range of the display device. Thus, the typical requirements of further post-processing such as non-linear stretching for display are often not necessary. This is preferred in computer graphics applications where the primary need is to provide a good visual display.

- **S-II**: The earlier strategy (**S-I**), although suits very well for pure display applications, it does significantly alter the relationship between the average intensity of the input hyperspectral image and the output fused image. One may expect the average intensity values of the input and the output to remain the same, in which case the fusion is said to preserve the radiometric mean of the data. This preservation could be of importance in some remote sensing applications. The radiometric fidelity facilitates a comparison and visual inspection of two different datasets which is not possible with the earlier mapping strategy. To achieve this goal, we would like to map the pixels from the input toward the mean gray value of the input data (i.e., hyperspectral data cube). This mapping helps maintaining the average intensity level of the output close to that of the input hyperspectral data. The fused image will have a low value of the relative bias which has been defined in [92] as the deviation of the mean of the fused image from the mean of the input hyperspectral bands. Thus, the fused image has a lesser degree of radiometric distortion, and is often preferred in remote sensing applications.

As our primary goal is to generate a fused image for visualization purposes, we follow strategy S-I. Thus, the fusion weights should be calculated in such a way that the gray values of the pixels of the fused image will be close to the central gray level. From an information theoretic point of view, this problem can be related to the problem of entropy maximization. We are addressing this problem for the mapping of the dynamic range of the hyperspectral scene across the wavelength spectrum to generate a single fused image. We define our objective function to compute a set of weights $\{\alpha_k(x, y)\}$ that will maximize the entropy ε_1 of the fused image $F(x, y)$ given by Eq. (7.2).

$$\varepsilon_1(\alpha) = -\int_x \int_y F(x, y) \log\left(\frac{F(x, y)}{0.50\,e}\right) dx\,dy, \tag{7.2}$$

where the factor of $0.50\,e$ in the denominator has been introduced to force the trivial solution to go to 0.50 as explained in S-I. It should be noted that in the present context, the entropy of an image has been defined over the normalized pixel values of the hyperspectral data. This should not be confused with the general and more commonly used definition of the image entropy as a measure of average information content in the picture. The definition of entropy as in Eq. (7.2) is commonly used in restoring astrophysical images under the terminology maximum entropy reconstruction [80]. If one wants to implement the algorithm for strategy (S-II), one should replace the denominator in Eq. (7.2) (i.e., $0.50\,e$) by m_Ie, which is the mean of the entire hyperspectral data I.

Let us take a look at the trivial solution for the maximization of Eq. (7.2), that is the solution one obtains after the iterative system converges. In this case, the system converges to provide a solution that is constant, i.e., $F(x, y) = 0.50 \ \forall (x, y)$. Although, a practical fusion system may not reach an exactly constant solution, it definitely indicates that the output image has a poor contrast. Thus, the maximization of entropy alone is not enough to obtain a sharply fused image. Therefore, it is essential to incorporate a complementary objective to the optimization function that would provide the necessary balance between the entropy maximization and the contrast of the fused image F.

We have already discussed the variance of an image as one of the contrast measures. A high value of variance indicates a well spread-out values of intensities in the fused image over the given range, yielding a visually sharper image. A small value of the variance indicates lack of details in the fused image. An excessive contrast leads to a non-natural appearance of the image, and thus, a moderate amount of contrast is necessary to obtain a sharp and visually appealing image. We incorporate an additional term (ε_2) in the objective function for maximizing the variance to produce high contrast resultant fused images.

$$\varepsilon_2(\alpha) = \frac{1}{XY} \int\limits_x \int\limits_y \left(F(x, y) - \frac{\left(\int_x \int_y F(x, y)\mathrm{d}x\,\mathrm{d}y \right)}{XY} \right)^2 \mathrm{d}x\,\mathrm{d}y. \qquad (7.3)$$

It can be seen that these two objectives (ε_1 and ε_2) are complementary in nature. The first criterion *pulls in* the far distinct pixels towards the mean gray level m_1. The second criteria *pushes out* the pixels away from the mean. The relative weightages of these two objectives decide how strong the particular objective is. It converges to the particular solution where both objectives have been balanced. That is, the resultant fused image possesses both the characteristics of a higher entropy and a higher local variance. The right combination of well-exposedness and contrast is what makes an image visually more appealing.

So far we have not considered any spatial correlation among the pixels while fusing them. The adjacent pixels in the hyperspectral data generally belong to the same or similar underlying objects in the scene which have similar material composition. Thus, such pixels in the scene exhibit a high degree of correlation with the neighborhood pixels. A simple and commonly used method to acknowledge this intra-band correlation is the inclusion of a certain kind of smoothness constraint in the fusion expression [145]. However, the images are discontinuous at edges and boundaries. Enforcing a smoothness constraint on the resultant fused image often leads to an excessive smoothing, thereby blurring edges and washing away the weak and minor features. This often results in smearing at edges producing visible artifacts in the fused image. Therefore, enforcing a smoothness constraint on the fused image not only deteriorates the quality of the result, but also contradicts to one of the objectives of obtaining high contrast images from Eq. (7.3). We want to acknowledge the strong spatial correlation among the input pixels, but we also want to avoid any smoothing of fusion result, especially at edges and boundaries. To achieve this, we incorporate an additional penalty term (ε_3) in the cost function which enforces a smoothness of the fusion weights (i.e., the α-matte) rather than in the fused image F. When the data in the input hyperspectral bands are smooth, one would expect the corresponding weights also to be smooth i.e., α_k should be smooth along the x and y directions. However, when contents of hyperspectral bands exhibit some degree of discontinuity, one would like the corresponding features to get an appropriate representation in the fused image as well. As we are not imposing any kind of smoothness constraint over the output, the features in the input data, representing the discontinuity (e.g., edges, boundaries), get preserved in the output which is formed through a linear combination of the input hyperspectral bands. Thus, if the input bands have a good contrast, it does not get deteriorated during fusion. The smoothness objective for the α-matte can be written as,

$$\varepsilon_3(\alpha) = \int\limits_x \int\limits_y \sum_{k=1}^{K} \left(\alpha_{k_x}^2(x, y) + \alpha_{k_y}^2(x, y) \right) \, dx \, dy, \qquad (7.4)$$

where α_{k_x} and α_{k_y} denote derivatives of the fusion weight α_k in the x- and y-directions, respectively.

Now, let us formulate the overall objective function for a constrained minimization by combining Eqs. (7.2)–(7.4) as shown in Eq. (7.5).

$$J(\alpha) = -\varepsilon_1(\alpha) - \lambda_v \varepsilon_2(\alpha) + \lambda_s \varepsilon_3(\alpha) \qquad (7.5)$$

$$\text{subject to, } \alpha_k(x, y) \geq 0 \quad \forall k, \text{ and } \sum_{k=1}^{K} \alpha_k(x, y) = 1 \quad \forall (x, y), \qquad (7.6)$$

where λ_v and λ_s are the regularization parameters that define the weightage given to the variance term, and the smoothness term as compared to the entropy term, respectively. From Eq. (7.5), we can infer that the fused image can be obtained by solving the problem of calculus of variation.

7.3.2 Variational Solution

We have discussed two constraints on the fusion weights, viz., unity norm and non-negativity. The first constraint refers to the sum of all the weights across all spectral observations at a given spatial location. This value should be unity which indicates the relative contribution of all the observations toward the final combination. The normalization constraint can easily be incorporated into the cost function with the help of a Lagrangian multiplier. An explicit term enforcing the sum of the fusion weights to be one can be easily augmented with the existing cost function, as it will be shown later.

An addition of the non-negativity constraint on the fusion weights, however, converts the problem into a computationally demanding constrained optimization problem. We need a computationally simple solution without sacrificing on the positivity of the fusion weights. We show how to accomplish this task with the use of an auxiliary variable. Here we introduce a set of auxiliary variables w, and define it as the positive square root of the matte α. Thus the original weights α can be replaced as,

$$\alpha_k(x, y) \triangleq w_k^2(x, y) \quad \forall (x, y). \qquad (7.7)$$

Here, we need to modify the cost function and the constraints in Eq. (7.5) appropriately in terms of w as the following. The expressions for entropy and variance terms are modified by replacing α by w^2. In case of the smoothness term, however, the smoothness in w_k which is a positive square root of w_k^2 also implies smoothness in the actual weights w_k^2. Therefore, we can as well impose the smoothness in w_k as an explicit constraint to ensure the smoothness in the actual weights w_k^2. It can be easily observed that the actual weights α_k (i.e., w_k^2) are always non-negative, irrespective of the sign of w. The constraint for unity norm, however, should be explicitly added to ensure that for every pixel (x, y), the sum of the weights should equal 1. After modifying the weights to the square term, the unity norm constraint can be written as,

$$\sum_{k=1}^{K} w_k^2(x, y) = 1. \tag{7.8}$$

Consider the constraint given in Eq. (7.5) which specifies the condition on the fusion weights α_k. The summation constraint indicates that the fusion weights for a given observation lie on the hyperplane $\sum \alpha_k = 1$ in the first quadrant. The constraint given by the auxiliary variables specifies that the fusion weights in the form of auxiliary variables should lie on a unit hyper-sphere. It will be seen later that such a unity norm constraint can be easily enforced externally while solving Eq. (7.5) as an unconstrained optimization problem.

Now we provide the solution to the multi-objective optimization problem using the Euler-Lagrange equation that we have already discussed in the previous chapter. Since we are mostly dealing with the 1-D spectral array at a given location in a 2-D spatial plane, instead of working with 2-D bands I_k, $k = 1, 2, \cdots, K$, it is more convenient to work with the spectral array notation as explained earlier in Chap. 2. The spectral array or the vector at location (x, y) has been referred to as $\mathbf{s}(x, y)$, where $\mathbf{s} \in \mathbb{R}^K$. The k-th element of this vector is denoted by $s_k(x, y)$. It should be, however, noted that the change of notation is purely meant for the ease of understanding, and to avoid any possible confusion when the same data is being referred from different dimensions. The entire hyperspectral image as a cube can be denoted by $\mathbf{I}$. We shall denote the weight vector (in w) at the spatial location (x, y) by $\mathbf{w}^2(x, y)$, where $\mathbf{w}^2(x, y) \in \mathbb{R}^K$. This vector represents an element-wise product of the vector $\mathbf{w}(x, y) \in \mathbb{R}^K$ with itself. This is also known as the *Hadamard product*. Mathematically, we can write this expression by Eq. (7.9).

$$\mathbf{w}^2(x, y) = \{w_k(x, y)\, w_k(x, y)\ \forall k\} = \mathbf{w}(x, y) \circ \mathbf{w}(x, y) \tag{7.9}$$

where $\circ$ represents an element-wise product operator. Using the vector notation, the resultant fused image $F(x, y)$ can be represented in the form of a dot product of the input data vector $\mathbf{s}(x, y)$ at the corresponding pixel location (x, y) with the weight vector $\mathbf{w}^2(x, y)$ at the same spatial location, i.e.,

$$F(x, y) = \mathbf{s}(x, y).\mathbf{w}^2(x, y) = \mathbf{s}^T(x, y)\, \mathbf{w}^2(x, y). \tag{7.10}$$

The combination of Eqs. (7.5) and (7.8), gives the following cost functional,

$$J(\mathbf{w}) = \int_x \int_y \left\{ \left(\mathbf{s}^T \mathbf{w}^2\right) \log \left(\frac{\mathbf{s}^T \mathbf{w}^2}{0.50\,e}\right) - \lambda_v \left(\mathbf{s}^T \mathbf{w}^2 - \frac{1}{XY} \left(\int_x \int_y \mathbf{s}^T \mathbf{w}^2 dx\, dy\right)\right)^2 \right.$$

$$\left. + \lambda_s \left(\|\mathbf{w}_x\|^2 + \|\mathbf{w}_y\|^2\right) + \mu \left(\mathbf{w}^T \mathbf{w} - 1\right) \right\} dx\, dy. \tag{7.11}$$

Here λ_v and λ_s are the regularization parameters, while μ is the Lagrangian multiplier for the unity norm constraint. The arguments x and y of the functions are omitted for the purpose of brevity ($\mathbf{s} \equiv \mathbf{s}(x, y)$, etc.). The solution of Eq. (7.11) is obtained using the corresponding Euler-Lagrange equation,

$$\frac{\partial J_I}{\partial w} - \frac{\partial}{\partial x}\left(\frac{\partial J_I}{\partial w_x}\right) - \frac{\partial}{\partial y}\left(\frac{\partial J_I}{\partial w_y}\right) = 0, \tag{7.12}$$

where $J_I(w, w_x, w_y)$ is the integrand in Eq. (7.11). On simplification, Eq. (7.12) becomes,

$$(\mathbf{s} \circ \mathbf{w})\left(1 + \log \left(\mathbf{s}^T \mathbf{w}^2\right) - \log(0.5e)\right) - 2\lambda_v \left(\mathbf{s}^T \mathbf{w}^2 - \frac{\int_x \int_y \mathbf{s}^T \mathbf{w}^2 dx\, dy}{XY}\right)$$

$$\left(\mathbf{s} \circ \mathbf{w} - \frac{\int_x \int_y \mathbf{s} \circ \mathbf{w}\, dx\, dy}{XY}\right) - \lambda_s \nabla^2 \mathbf{w} + \mu \mathbf{w} = \mathbf{0}, \tag{7.13}$$

where $\circ$ and ∇^2 represent the element-wise product operator, and the Laplacian operator, respectively. For the RHS of the equation, $\mathbf{0}$ indicates a zero vector. The Laplacian operator for a 2-D function w is given by Eq. (7.14).

$$\nabla^2 w = \frac{\partial^2 w}{\partial x^2} + \frac{\partial^2 w}{\partial y^2}. \tag{7.14}$$

A discrete approximation of the Laplacian operator is given as [76]

$$\nabla^2 \mathbf{w}(x, y) \approx \frac{4}{\delta^2}(\bar{\mathbf{w}}(x, y) - \mathbf{w}(x, y)),$$

where $\bar{\mathbf{w}}(x, y)$ represents the local average of the weight vectors in the X and Y dimensions, and δ is the distance between adjacent pixels, trivially set to 1. After discretization, the iterative solution for the weight vector $\mathbf{w}$ can be obtained by re-arranging the terms as given by Eq. (7.15).

$$
\mathbf{w}^{(m+1)} = \frac{1}{1+\frac{\mu\delta^2}{4\lambda_s}} \left\{ \bar{\mathbf{w}}^{(m)} - \frac{\delta^2}{4\lambda_s} \left(\mathbf{s} \circ \mathbf{w}^{(m)} \left(1 + \log\left(\mathbf{s}^T \mathbf{w}^{(m)2}\right) \right. \right. \right.
$$

$$
- \log(0.5e) - 2\lambda_v \left(\mathbf{s}^T \mathbf{w}^{(m)2} - \frac{\sum_x \sum_y \mathbf{s}^T \mathbf{w}^{(m)2}}{XY} \right) \right)
$$

$$
\left. \left. +2\frac{\lambda_v}{XY} \left(\mathbf{s}^T \mathbf{w}^{(m)} - \frac{\sum_x \sum_y \mathbf{s}^T \mathbf{w}^{(m)2}}{XY} \right) \sum_x \sum_y \mathbf{s} \circ \mathbf{w}^{(m)} \right) \right\}, \quad (7.15)
$$

where (m) is the index of iteration. The scalar μ appears only as a part of a positive scaling factor in Eq. (7.15). Also, the purpose of μ is only to enforce the unit length of the weight vector. If we want to avoid this scaling factor, we have to explicitly normalize the weight vector $\mathbf{w}^{(m+1)}$ at each iteration to satisfy the necessary constraint given in Eq. (7.8) [76]. Here we introduce an intermediate variable z to represent un-normalized weights, w. The final solution is thus given by Eq. (7.16).

$$
\mathbf{z}^{(m+1)} = \bar{\mathbf{w}}^{(m)} - \frac{\delta^2}{4\lambda_s} \left(\mathbf{s} \circ \mathbf{w}^{(m)} \left(1 + \log\left(\mathbf{s}^T \mathbf{w}^{(m)2}\right) - \log(0.5e) \right. \right.
$$

$$
-2\lambda_v \left(\mathbf{s}^T \mathbf{w}^{(m)2} - \frac{\sum_x \sum_y \mathbf{s}^T \mathbf{w}^{(m)2}}{XY} \right) \right)
$$

$$
\left. +2\frac{\lambda_v}{XY} \left(\mathbf{s}^T \mathbf{w}^{(m)2} - \frac{\sum_x \sum_y \mathbf{s}^T \mathbf{w}^{(m)2}}{XY} \right) \sum_x \sum_y \mathbf{s} \circ \mathbf{w}^{(m)} \right) \quad (7.16)
$$

$$
\mathbf{w}^{(m+1)} = +\sqrt{\frac{\mathbf{z}^{(m+1)} \circ \mathbf{z}^{(m+1)}}{\mathbf{z}^{(m+1)T} \mathbf{z}^{(m+1)}}}. \quad (7.17)
$$

The above equation provides a solution of the unconstrained optimization problem to solve fusion problem. As explained earlier, the resultant fused image is formed by linear weighted combination of the input hyperspectral bands, while the fusion weights have been computed using the aforementioned unconstrained optimization process. The basic process of fusion is, thus, provided by Eq. (7.1) with $\alpha_k(x, y) = w_k^2(x, y)$. We expect the fused image to be centered around the mean radiometric value of the data cube $\mathbf{I}$, and to have a high contrast. The estimated α-matte is locally smooth, but not necessarily the fused image.

7.4 Implementation

The solution presented in this chapter requires two regularization parameters that define the relative weightage for each of the corresponding objectives. These weights essentially determine the nature of the fused image in terms of the relative strength of the desired qualities, and the rate of convergence of the solution. The values of these parameters, thus, play an important role in the process of fusion. The selection of regularization parameters is known to be a difficult problem which is typically solved using a cross-validation technique [64]. However, fortunately, the final result of fusion is not very sensitive to the exact value of λ_v, but depends on the order of the value, i.e., $\lambda_v = 1, 10, 100, 1000, \cdots$. We have used the values in the range of 10^2 which have been found to provide a good balance among the competing terms in the objective function.

The value of λ_s should be less than λ_v, as λ_s serves as the relative weightage given to the smoothness term of the minimization functional. It should be noted that the smoothness penalty should not be very strong, as it may produce almost similar values of fusion weights for neighboring pixels. Thus, a high value of this term may reduce contrast in the fused image which would lead to spectral averaging of the image bands. We have selected this value to be 5–10 % of the regularization weight assigned to the variance term.

Several strategies to stop the iterative minimization process may be employed. We have followed the commonly used relative cost based criteria to conclude the iterative procedure as the one employed in the previous chapters of this monograph as well. During this procedure, the total cost of the functional $J^{(m)}$ is computed after every iteration (m). The change in the value of the functional over the successive iterations is calculated, i.e., $\nabla J^{(m)} = J^{(m)} - J^{(m-1)}$, $m \geq 1$. The stopping rule is defined in terms of this relative difference of the cost functional, i.e., $\frac{\nabla J^{(m)}}{J^{(m)}}$. It was seen that typically the fusion process took 8–10 iterations to converge.

7.5 Experimental Results

In order to maintain the uniformity across chapters, we have used the same two datasets for the demonstration of this fusion technique. More results are provided in Chap. 10.

The urban hyperspectral data consist of 242 bands with dimensions (512×256) each. To generate a single grayscale fused image, we have processed the entire data cube at once, while to produce an RGB output, we have partitioned the data into 3 subsets. These subsets undergo fusion independently to generate three images which are then assigned to the red, green, and blue channels to provide a fused RGB image. The assignment of colors is not directly related to the actual wavelengths of these primary colors, and hence, several pseudo-color schemes may be used to present the result in an enhanced manner. Figure 7.1a shows the result of combining all bands using the optimization-based solution. This result represents fusion over the spectral bandwidth of nearly 2200 nm. An RGB version of the fused image is shown in Fig. 7.1b which is a result of assignment of pseudo colors to the resultants of fusion of nearly one-third data each. The results of fusing the moffett$_2$ dataset have been provided in Fig. 7.2. While the figure on the left provides a grayscale version of the

(a)   **(b)**

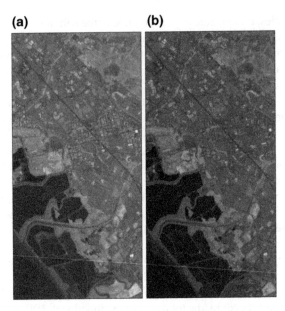

Fig. 7.1 Results of the optimization-based fusion of the urban image cube from the Hyperion. **a** Grayscale fused image, and **b** RGB fused image (©2012 IEEE, Ref: [90])

(a)   **(b)**

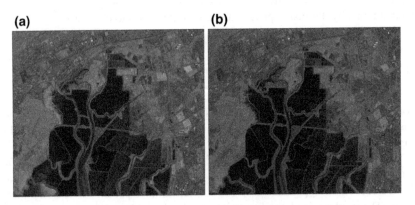

Fig. 7.2 Results of the optimization-based fusion of the moffett$_2$ image cube from the AVIRIS. **a** Grayscale fused image, and **b** RGB fused image (©2012 IEEE, Ref: [90])

output, it makes the features in the data quite apparent with a high value of contrast. An RGB version of the output is shown in Fig. 7.2b. This image has been generated by three independently fused non-overlapping subsets of the moffett$_2$ data. In both the cases, we have demonstrated the RGB results using a simple strategy—partition the input data sequentially along the wavelength into three groups having nearly an equal number of bands. Contiguity of wavelength during visualization is important as all these bands are clubbed into a single color channel. Each of the groups have been fused separately, and the three fused images are assigned to R, G, and B channels to form a composite color output.

7.6 Summary

This chapter develops a fusion methodology that transforms the problem into an optimization framework where visualization is the primary aim. Therefore, it focuses on computing the fusion weights optimally in order to provide better visualization results. The novelty of the discussed solution lies in the fact that the fusion process is driven by the desired properties of the output image as opposed to the traditional methods which are driven by the input. We have explained a multi-objective cost function based on some of these image properties, and then discussed the solution using the Euler-Lagrange equation. Being completely output-driven, this solution does not require any assumption on how the weights should be related to the input images.

Our discussion also includes all typical constraints of pixel-based image fusion. We have, thus, demonstrated how fusion can be represented as an unconstrained optimization problem by introducing an auxiliary variable which leads to a computationally simpler and easier solution.

Chapter 8
Band Selection: Revisited

8.1 Introduction

In last few chapters we have seen four different methodologies for fusion of hyperspectral images. Chapters 3, and 5 describe fusion techniques based on extracting some of the salient characteristics of the input data. In the last chapter, we have discussed alternately how the fusion technique can be based on the desired characteristics of the output. For most of the users and analysts, the visual quality of the fused image is of primary importance. The key issue is then how to efficiently capture the features from the input images and transfer them into saliency of the fused image. These features are mainly application dependent, and thus, the technique discussed in Chap. 7 is more of a generic nature. The idea behind the optimization-based fusion technique described in the last chapter is to develop a fusion technique that *explicitly* deals with the properties of the resultant output image (to be generated through the fusion process).

We have also discussed a method of band selection in Chap. 4 where a specific subset of hyperspectral bands was selected from the input image based on the conditional entropy measure. We have also observed that one can achieve almost a similar fusion output by using a small fraction of hyperspectral data. If we choose the bands properly, the resultant fused image obtained from this subset of the data provides a very little degradation in the quality of the output when compared with the resultant image obtained from the fusion of the entire dataset using the same fusion technique. The remaining (discarded) bands, thus, contribute a very little amount of information towards fusion as most of the independent information has already been captured. In order to facilitate this band selection, the conditional entropy of a band of the hyperspectral data given the set of already selected bands has been calculated.

The idea of the output-based fusion motivates us to develop an alternate band selection method that selects a subset of bands depending upon whether the fused image obtained by inclusion of a particular band to the subset will be significantly different from the fused image obtained without inclusion of that band to the subset. In this chapter, we shall again discuss the band selection method, but from an

S. Chaudhuri and K. Kotwal, *Hyperspectral Image Fusion*,
DOI: 10.1007/978-1-4614-7470-8_8, © Springer Science+Business Media New York 2013

output-based perspective. We would again like to select only a few specific bands to accomplish an efficient fusion of hyperspectral images without much sacrificing the quality of the fusion output. We discuss an information theoretic strategy for the selection of specific image bands of the hyperspectral data cube using the corresponding intermediate output of the fusion process. It should be noted that the input-based band selection scheme is independent of the fusion technique. The present output-based band selection, however, makes use of the intermediate outputs of the fusion process, and thus, is very much dependent on the fusion technique employed.

The conditional entropy acts as a measure of the amount of additional information contained in the band. The rationale and theory behind the use of conditional entropy measure has already been explained in Chap. 4. The next section, Sect. 8.2 explores the possibility of having an output-based scheme of band selection. Section 8.3 consists of some of the experimental results and the performance analysis of the output-based band selection scheme. Section 8.4 presents the summary.

8.2 Output-Based Band Selection

We have discussed an input-based scheme of band selection, and its special case for a spectrally ordered dataset in Chap. 4. This scheme aims at removing the redundancy in the input prior to the actual fusion. This scheme is independent of the fusion technique, and hence, does not consider how the fused image gets generated from the selected image bands. Different fusion techniques have varying capabilities in terms of combining information from two different sources into a single representation. Thus, as more and more images are fused, one cannot prove that an increase in the information content in the fused image is always either equal or proportional to the conditional entropy of the additional image being fused with. As one is mainly interested in the final result of fusion for visualization purposes, one may expect to choose a specific subset of image bands which produce a fused image of a higher quality for the given fusion technique.

Let us consider a hyperspectral image $\mathbf{I}$ containing K bands. As usual, we can represent this image as a set of K bands, $\{I_k; k = 1, 2, \ldots, K\}$. Let F_p, $p = 1, 2, \ldots$ be the resultant fused image from the fusion of p selected bands using a pixel-based fusion rule $\mathscr{F}$. The next band to be selected for fusion using $\mathscr{F}$ should possess a significant amount of additional information compared to the existing fusion output, i.e., F_p, for the fusion to be statistically efficient. We explain an output-based scheme for the selection of bands which evaluates the redundancy of the successive input image bands as compared to the corresponding intermediate output image using the conditional entropy, and selects only the less redundant bands for fusion.

Initially, the first band is selected trivially for fusion, and it is considered to be equivalent to the first intermediate fused image F_1. The conditional entropy of the successive input bands with respect to this intermediate resultant image is evaluated. The next band is selected when the corresponding conditional entropy exceeds a pre-determined threshold, i.e., when the corresponding band possesses a sufficiently high amount of additional information than the presently available fused image. The threshold, as before, is set to an appropriate fraction of the entropy of the band under consideration. Once the band is selected, a new fused image is formed by combining all the selected bands using the fusion technique $\mathscr{F}$. Subsequently, for every additional image band, the conditional entropy with respect to the corresponding fused image obtained by combining all of the previously selected bands is calculated. Thus, given a set of hyperspectral images $\{I_k; k = 1, 2, \ldots, K\}$, the p-th image band selected for fusion is given by Eq. (8.1).

$$p = \arg\min_r \{H(I_r|F_{r-1}) \geq \theta'\}, \tag{8.1}$$

where $H(I_r|F_{r-1})$ represents the entropy of the image I_r conditioned on the fused image F_{r-1}, obtained from all of the already selected bands up to I_{r-1} using the same fusion technique $\mathscr{F}$. It should be noted that the actual number of bands undergoing fusion are much less than $(r - 1)$. The threshold θ' is chosen again as a suitable fraction κ of $H(I_r)$,

$$\theta' = \kappa\, H(I_r), \qquad 0 < \kappa < 1. \tag{8.2}$$

This procedure is continued until the entire dataset is exhausted. This scheme exploits the redundancy in the input data with respect to the intermediate output, as opposed to purely input-based band selection. Hence this is an example of *fusion process-in-the-loop* technique for band selection. The resultant fused image contains most of the features of the entire data as it rejects an image band when it is highly similar to the existing output at the corresponding fusion stage.

The methodology of the selection of the bands based on the output is thus directly dependent upon the chosen fusion technique. Several pixel-based techniques for fusion of hyperspectral images have been proposed [71, 79, 88, 189]. Different fusion techniques measure the saliency of the pixels in different ways, and assign suitable weights to them. The resultant fused images are generated by a linear combination of the pixels across all the bands. During the process of fusion, different fusion rules introduce different amounts of loss and noise, and thus, an *a priori* estimation of the entropy of the output is a difficult task. Further, due to the dependencies on the fusion technique, it is difficult to device a generalized model for $H(I_r|F_{r-1})$ and hence to estimate the savings in computation as was done for the input-based band selection. Similar to the input-based selection of bands, this type of output-based selection is also based on the greedy technique as one cannot undo the selection of any band.

At this point one may point out that ideally one should fuse a candidate band first and then we should compute the conditional entropy of the newly fused image with respect to the previous fusion result, and if this measure exceeds a threshold then the candidate band should be selected. While this sounds logical, this is of no practical value as the majority of the computational requirement is consumed by the fusion process and not the computation of conditional entropy. Thus, if the fusion has already been carried out, there is no reason why the candidate band should then be discarded.

8.3 Experimental Results

In this section, we discuss some experimental results for the output-based band selection scheme over the two datasets- the urban data, and the moffett$_2$ data which the readers are familiar with. The performance of this scheme is dependent on the fusion technique employed. Although, our motivation of developing the output-based band selection comes from the output-based fusion scheme presented in the last chapter, we employ the bilateral filtering-based fusion technique (Chap. 3) for the illustration purposes. There are two reasons for the choice of fusion technique. First, we have provided the experimental results for the input-based band selection scheme using the same bilateral filtering-based fusion technique in Chap. 4. We have also analyzed the performance of the band selection scheme using various measures over the output images fused using the bilateral filtering-based technique. Thus, providing the resultant images using the same fusion technique over the bands selected using the output-based scheme will facilitate the readers to compare the performances of both the band selection schemes. Second, the output-based band selection requires generation of the so-called intermediate fusion output for every additional band selected. The bilateral filtering-based fusion technique is a non-iterative, faster process as opposed to the optimization-based fusion which is iterative in nature. Thus, the bilateral filtering-based fusion technique turns out to be a quick, yet reliable option for the analysis and demonstration purposes.

Figure 8.1a shows the result of fusion of the urban data by the Hyperion using the bilateral filtering-based technique. The resultant image is, however, obtained from the combining of only 27 bands out of a total of nearly 170 useful bands in the original data. One may compare this figure to Fig. 8.1b representing fusion of the entire urban data cube. This figure is essentially the same as the corresponding fused image in Chap. 4 as the fused technique and the set of bands being fused are exactly the same. It may be observed that Fig. 8.1a brings out most of the features as does the other image, and does not introduce any visible artifacts in the result. Thus, the resultant images fused over the subsets of the hyperspectral data selected using the output-based scheme are comparable to the resultant images produced from fusion of the entire dataset using the same fusion technique in terms of visual quality.

As we have already introduced two parameters for the performance evaluation of fused images *viz.*, the entropy, and the average gradient, we shall continue to use the same in the present chapter. More details on evaluation of fusion techniques will be provided in the next chapter, Chap. 9.

The entropy represents the information content in the image. Figure 8.2a shows the entropies of the fused images for different values of the threshold parameter κ. As κ increases, the number of bands selected for fusion reduces, and thus, one may expect smaller values of the entropy as fewer input bands are contributing towards

Fig. 8.1 Results of fusion of the Hyperion data applied over a subset of bands selected using the output-based scheme. **a** shows the results of fusion of only 27 selected bands when κ was set to 0.70. **b** shows the corresponding result of fusion of the entire data

(a)   **(b)**

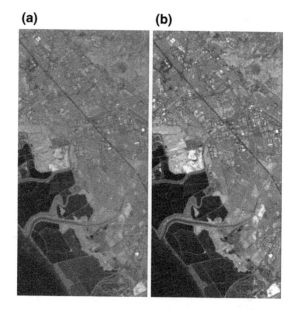

the fusion output. However, it can be observed from the plots that the performance remains almost the same till the value of κ becomes smaller than 0.60, and it drops marginally beyond the value of 0.60 signifying that one may discard a large chunk of bands in order to make fusion process computationally efficient, and yet not lose on the information content in the fused image.

The average gradient $\bar{g}$ of an image evaluates the quality in terms of the spatial gradient of the image at all pixels. Figure 8.2b represents the relation between the average gradient ($\bar{g}$) with respect to the variation in the threshold parameter κ. For the values of κ nearly up to 0.40, the band selection scheme does not discard any of the bands from both of the hyperspectral data. This suggests that for any candidate band, the conditional entropy with respect to the fused image is at least greater than 40 % of its own entropy. Therefore, the corresponding plots of the average gradient are constant for $\kappa = 0$ to 0.40. A gradual decrease in the values of the gradient can be observed for higher values of κ as fewer bands get selected resulting in a slight degradation in the quality of the fused image. This reduction in the entropy and gradient values is very minimal. For example, for $\kappa = 0.80$, the output-based scheme selects less than $1/10$-th of the hyperspectral data, and yet the reduction in the values of the evaluation measures is less than 10 %. The nature of entropy and average gradient plots from Figs. 8.2a, b is similar to that of the plots of these measures for the input-based band selection scheme as one can observe from Figs. (4.4a, b), respectively. It may be understood that both these schemes suggest that one may carry out efficient fusion using only a fraction of bands from the hyperspectral image.

While maintaining the consistency with the performance evaluation procedure for the input-based band selection discussed in Chap. 4, we shall proceed with the same

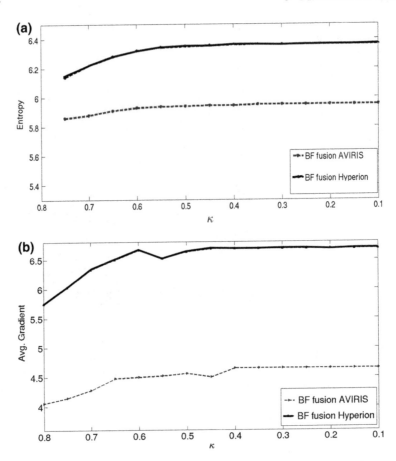

Fig. 8.2 Performance evaluation of the scheme of output-based band selection using different measures for various values of κ fused using the bilateral filtering-based technique. **a** entropy measure, and **b** average gradient

analysis using the PSNR measure. This measure enables us to understand the change or the difference between the output image fused using the subset of selected bands and a reference image in terms of the mean square error. The reference image, as usual, is the image one would obtain from fusion of the entire dataset. The output-based band selection being fusion technique dependent, it is essential to carry out fusion procedure using the same technique, which in this case is the bilateral filtering-based technique. We would like to analyze the performance of band selection by varying the threshold parameter κ against the changes in the PSNR of the fused image thus obtained. Since the reference image is obtained from fusion of the entire dataset, it is equivalent to setting κ to be zero. Figure 8.3 indicates the nature of the radiometric error for the fused image as κ varies from 0 to 1.0. The output-based scheme selects all the bands for the values of κ upto 0.40, and therefore, the resultant is the same as

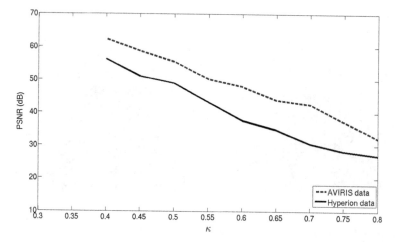

Fig. 8.3 Plots of PSNR of the fused images for various values of κ on the Hyperion and the AVIRIS data using the bilateral filtering-based fusion technique over the subsets selected using the output-based scheme. The resultant image from fusion of the entire dataset using the same technique is used as a reference in each case

the reference image. Similar to the input-based band selection, the values of PSNR drop very gradually when κ is increased. For $\kappa = 0.65$, this scheme selected 71 bands of the urban hyperspectral image which is nearly $^1/_3$-rd the entire dataset, however, the fused image provides a PSNR above 30 dB which indicates a good visual match between the two images. For the same value of κ, 110 bands of the moffett$_2$ data were selected, yielding the corresponding fused image with a PSNR of nearly 43 dB. A PSNR above 40 dB can be observed in the resultant image generated from fusion of only 12 bands when the threshold parameter κ is set to 0.70 in the case of the moffett$_2$ hyperspectral image. This result compares very well with that for the input-based band selection scheme shown in Fig. 4.5. Therefore, both the schemes of band selection provide a subset of bands which is capable of producing comparable results with only a fraction of the data. Similar characteristic plots have been observed for other pixel based fusion techniques also.

Although visual comparison is a subjective measure, it is often important, especially when the primary purpose of fusion is the visualization of scene contents by a human analyst. We provide some of the output images resulting from fusion of subsets of the moffett$_2$ hyperspectral image where the bands have been selected using the output-based scheme. The fusion results of the output-based band selection are provided in Fig. 8.4. The images in Figs. 8.4a–d are obtained from the subsets of the moffett$_2$ data where the cardinality varies with relation to the variation in the threshold parameter κ. Figure 8.4a shows the result of fusion of the entire dataset, and thus it acts as the reference for the evaluation purpose. Figure 8.4b is the result of fusion of 142 bands by setting κ to be 0.50. As for the values of $0 \leq \kappa \leq 0.40$, the band selection scheme does not discard any of the input bands, and hence the fused images obtained for the values of κ less than 0.40, are essentially the same

(a)   **(b)**

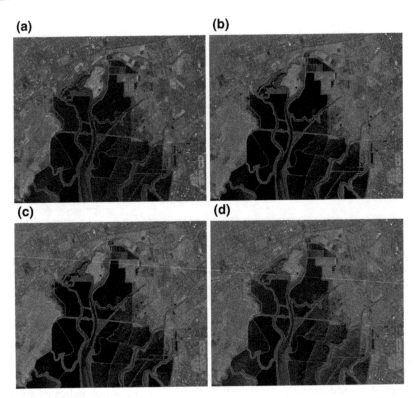

(c)   **(d)**

Fig. 8.4 Results of bilateral filtering-based fusion of the moffett$_2$ data from the AVIRIS applied over a subset of bands selected using the output-based scheme. **a–d** show the results of fusion for κ values 0, 0.50, 0.65 and 0.75, respectively

as the reference fused image. The result of fusion of 110 bands of the moffett$_2$ data is shown in Fig. 8.4c where the threshold parameter κ was set to 0.65. Finally, the resultant of fusion of 14 bands which are nearly $^1/_{10}$-th of the hyperspectral image is shown in Fig. 8.4d. These bands were obtained by choosing κ to be 0.75, and yet, the fused image provides a PSNR of around 35 dB.

The Bhattacharyya distance is another measure for evaluation of the closeness between two images. Our interest lies in evaluating the closeness between the fused images obtained from the subset of the bands obtained from the output-based selection scheme, and the resultant image from fusion of the entire dataset, designated as the reference image throughout our discussion. The Bhattacharyya coefficient is a distance metric that indicates the degree of overlap between the histograms of two images. The plot of the Bhattacharyya coefficient (BC) between the reference image and the resultant images obtained from fusion of subset of bands selected using the output-based scheme is shown in Fig. 8.5. These subsets of hyperspectral bands are obtained by varying the threshold parameter κ, and subsequently fusing the bands using the bilateral filtering-based technique. The Bhattacharyya coefficient (BC) can be observed to gradually increase for increasing values of κ beyond 0.50.

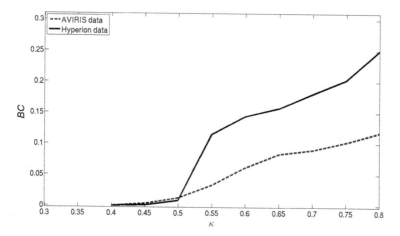

Fig. 8.5 Plots of Bhattacharyya coefficient between the resultant image from the fusion of entire data and the resultant images for various values of κ for the urban and the moffett$_2$ data using the bilateral filtering-based fusion technique over the subsets selected using the output-based band selection scheme

The computational requirement $\mathbf{W}(\kappa)$ for the output-based band selection scheme can be provided by Eq. (8.3) which is quite similar to the requirement for the input-based band selection scheme discussed in Chap. 4.

$$\mathbf{W}(\kappa) = \gamma\,\mathscr{B}(\kappa) + c_E, \tag{8.3}$$

where $\mathscr{B}(\kappa)$ is the number of bands selected for a given threshold κ, and c_E denotes the amount of computation for the evaluation of conditional entropies of the successive image bands. The γ factor represents the proportionality factor to account for computational requirements of a chosen fusion technique. The only difference lies in the fact that an intermediate fusion has been carried out first before the computation of the conditional entropies, unlike in the input-based scheme wherein fusion follows the entropy computation. This inherently assumes that one uses an incremental fusion technique for the output-based scheme. The incremental fusion makes use of the resultant of the previous intermediate fusion for carrying out the next fusion process. Thus, if $F_{(k-1)}$ is the intermediate fused image from fusion of upto $(k - 1)$ bands from the data using the fusion technique $\mathscr{F}$ (the actual number of bands being fused might be much less than $(k - 1)$ depending upon the values of κ), and I_k be the next input band to be selected for fusion, then the output of the fusion, i.e., F_k, is obtained as $F_k \equiv \mathscr{F}(F_{k-1}, I_k)$. Unfortunately, the incremental fusion methods may suffer from numerical inaccuracies, and one may have to recompute the fusion of all selected bands after every additional selection of a band. In this scenario, the computation increases after every selected band and one may not have any computational advantage over the normal scheme of fusion of all bands.

8.4 Summary

This chapter extends the discussion on information theoretic selection of specific
bands of hyperspectral image for efficient fusion. Similar to the previous band selection scheme discussed earlier in this monograph, the present scheme also makes
use of the conditional entropy as the measure of additional information. A band is
selected when it is less correlated with the corresponding intermediate output. Thus,
this scheme aims at the removal of the redundancy in the input hyperspectral data
depending on the contents of the fusion output. Quite naturally, the performance of
this scheme is dependent on the employed fusion technique, unlike in the input-based
band selection technique.

We have also discussed performance evaluation of the output-based band selection scheme using the bilateral filtering-based fusion technique. The performance
measures of the fusion results over two test datasets substantiate the efficiency of the
band selection scheme. A subjective evaluation of the results of these datasets is also
seen to be in agreement with the quantitative analysis.

Chapter 9
Performance Assessment of Fusion Techniques

9.1 Introduction

A quantitative analysis enables understanding of various pros and cons of a fusion technique. A human observer can judge the performance of a fusion technique in terms of the visual quality of the fused image. However, this approach is completely subjective, and thus may vary for different observers. Different subjects can perceive the same image differently, and their perception may be based on several psycho-visual factors rather than an objective assessment of the scene.

We have discussed in Chap. 2 the problems associated with the subjective assessment- it is both expensive and time consuming. Additionally, the assessment is likely to vary from person to person, and hence, it is not consistent. In order to obtain a unique assessment score, the image may be subjected to a large number of observers, and their individual scores are statistically averaged. This whole process, however, turns out to be tedious, time-consuming, and yet not very accurate. An objective assessment of the image quality alleviates most of these problems associated with the subjective quality assessment. In this process of analyzing the quality of an image, several performance measures which can be calculated either from the image alone, or with reference to some other image are employed. These measures can be computed without any human intervention, and they do not get affected by any psycho-visual, or individual differences. An objective assessment, thus, provides a consistent outcome which facilitates comparison of different images. These procedures can also be very easily automated as they do not require any inputs from the human analyst. The objective performance measures bring automation and uniformity to the process of assessment. Also, these measures may be used to optimize the performance of image and video processing systems by integrating their formal expressions into the system itself [184]. However, what makes this problem difficult is the fact that there is no universally accepted solution for the quality assessment of images in remote sensing. In case of hyperspectral image fusion the performance evaluation faces following challenges:

S. Chaudhuri and K. Kotwal, *Hyperspectral Image Fusion*,
DOI: 10.1007/978-1-4614-7470-8_9, © Springer Science+Business Media New York 2013

- unavailability of the ground truth in most cases,
- large data volume, and
- ambiguity in the quantification of *useful* data which is largely dependent on the context and application.

Several attempts to quantify the useful information from the point of visualization have been made, yet there is no standardization in this process. Furthermore, most of these measures have been developed for generalized image fusion where only a few images are to be fused. In such cases, mathematical formulation of performance measures may be easy and intuitive. One can easily interpret the physical meaning of quantities and terms involved. However, this may not be the case when the number of constituent images increases.

In this chapter, we extend some of these measures for an objective assessment of fusion of hyperspectral images. We also explain several modifications in the definitions of some of the existing measures in order to facilitate a better evaluation. The field of hyperspectral image fusion is not as mature as the field of generalized fusion, where the later is enriched with a large number of different methodologies. Hence, one may wish to experiment with extensions of these techniques toward the fusion of hyperspectral data. We also explain a notion of *consistency* of a fusion technique to understand the behavior of a given technique when it is applied over a progressively increasing sets of images. We believe that the consistency analysis will help in deciding the suitability of a particular technique toward fusion of a large number of images. Illustrations of the usage of these measures and the consistency analysis over some of the datasets and fusion techniques used in the previous chapters have also been provided.

We begin with a definition of the fusion consistency in Sect. 9.2. We then present the analysis of different quantitative measures and discuss several modifications in Sect. 9.3. The performance assessment of some of the recent techniques of hyperspectral image fusion, including the ones discussed in the monograph, using these modified measures has been provided in Sect. 9.4. Section 9.5 summarizes the chapter.

9.2 Consistency of a Fusion Technique

Let us consider fusion as a mathematical technique that determines how the pixels across multiple bands (or observations in the case of generalized image fusion) are combined. We would like to study the behavior of a fusion technique as a function of the number of images being fused. The focus is to understand how the fusion technique (a fusion rule, to be precise) reacts to the varying number of images being fused. Does the performance of a technique degrade with an increasing number of input images? Is the technique able to perform better with different implementation schemes? We shall try to answer these questions by developing a notion of *consistency* of a fusion technique.

If we consider a hyperspectral image of dimensions $(X \times Y \times K)$ as a set of K different bands, then we can form K subsets consisting of incrementally increasing number of bands such that the k-th subset contains hyperspectral bands up to I_k, i.e., $\{I_1, I_2, \ldots, I_k\}$. Each of the subsets can be fused using some fusion technique $\mathscr{F}$ to produce what we refer to as the incrementally fused image. Let F_k be the incrementally fused image from the subset $\{I_1, I_2, \ldots, I_k\}$ as given by Eq. (9.1).

$$F_k(x, y) = \mathscr{F}(I_1, I_2, \ldots, I_k) \quad k = 1 \text{ to } K. \tag{9.1}$$

This procedure generates a set of such K incrementally fused images $\{F_k, k = 1 \text{ to } K\}$. We want to analyze these incrementally fused images F_k's in terms of some of the objective quality measures.

We define the consistency of a fusion technique as an asymptotic behavior while dealing with an increasing number of images with respect to a given performance measure. Formally we define the *consistency* of a fusion technique as follows: [92]

Definition 9.1 Given a set of hyperspectral bands $\{I_k; k = 1, 2, \ldots, K\}$, let $\{F_k; k = 1, 2, \ldots, K\}$ represent a set of incrementally fused images given by $F_k = \mathscr{F}(I_1, I_2, \ldots, I_k)$ using a fusion technique $\mathscr{F}$. Then for a given performance measure g, the sequence $g(k) \equiv g(F_k)$, $k = 1 \text{ to } K$, represents an incremental behavior of the given fusion technique. We define the technique to be **consistent** with respect to the given measure g, if the sequence $g(k)$ possesses the following properties.

Property 9.1 The sequence $\{g(k); k = 1, 2, \ldots, K\}$ is monotonic. An addition of a new band during fusion must necessarily add more information to the resultant image.

Property 9.2 The sequence $\{g(k); k = 1, 2, \ldots, K\}$ is a bounded sequence. This value must be finite as $K \to \infty$. e.g., for entropy, the upper bound on the value is defined by the bitdepth of an image band.

From Properties (9.1) and (9.2), we can deduce that $\{g(k)\}$ is a *Cauchy sequence*, i.e., given $\varepsilon > 0$, however small, there are sufficiently large positive numbers m, n, such that $|g(m) - g(n)| < \varepsilon$. As one starts adding hyperspectral bands to the process of fusion, the initial few ones contribute significantly and build up the process. However, as more bands get fused, their contribution gradually decreases, and the corresponding incrementally fused images do not differ much. In other words, the property of the sequence getting progressively closer indicates that the additive contribution of individual image bands to the fusion process gradually becomes insignificant as more image bands are being added.

As we have already explained, there exists a strong spatial correlation among the consecutive bands of the hyperspectral image. This inter-band correlation leads to a large amount of redundancy in the data. We have also explored a couple of schemes for selection of a subset of few, but specific bands which capture most of the information content in the hyperspectral image. In Chaps. 4 and 8, we have discussed in detail how the additional information in the band can be utilized for developing band selection schemes. Due to this redundant nature of the hyperspectral data, a particular band provides only a small amount of additional information towards the final result. One would want a fusion technique to exploit this intrinsic relationship among bands while extracting the information from them. A good fusion technique should be capable of extracting and combining information across different bands in such a manner, that the sequence $g(k)$ for the corresponding performance measure is monotonic in nature. That is, a good fusion technique should satisfy Property (9.1).

In order to check the consistency one first needs to fix a performance measure. The set of incrementally fused images $\{F_k\}$, that involve fusion of a progressively increasing number of image bands using a technique $\mathscr{F}$, needs to be generated. The exact definition of $\mathscr{F}$ depends on the specific technique used and we provide some illustrations with some existing fusion techniques later in this chapter. During subsequent discussions, let us refer to the kth incrementally fused image as F_k, and the final fused image resulting from fusion of the entire data will be referred to as F_K, such that,

$$F_k|_{k=K} = \mathscr{F}(I_1, I_2, \ldots, I_K) = F.$$

Notwithstanding the incremental behavior, we refer to the final fused image as F instead of F_k for notational ease.

9.3 Performance and Consistency Analysis

Now let us analyze the behavior of some of the commonly used performance measures over the sets of incrementally fused images using fusion technique $\mathscr{F}$. We group the performance measures into three categories-

- No reference quality measures,
- Quality measures with an asymptotic reference, and
- Participatory performance measures.

As the name suggests, the performance measures in the first category can be employed directly over the corresponding fused images which could be either incremental or final, as they do not require any reference image to evaluate the quality. These are stand alone measures that are generally easy to calculate, intuitive, and computationally inexpensive. These measures include entropy, variance, etc., which have been traditionally employed for evaluation of fusion quality. As more and more bands are used, One would like these measures to improve until they saturate as per Property (9.2).

The second category refers to the performance measures with an asymptotic reference that study the progressive behavior of a technique with reference to the final fused image for a given hyperspectral data. These measures help us understand the progression of the fusion technique with reference to the particular performance measure as more and more bands are added to the process of fusion. This is particularly useful when one wants to extend a generalized fusion technique towards fusion of a large number of images, e.g., hyperspectral data. As mentioned earlier, image fusion literature is quite rich as this area is being actively researched for nearly last two decades. However, research in hyperspectral image fusion has begun recently. Naturally, one would like to extend the techniques of generalized image fusion for the combining of bands of the hyperspectral data. However, one needs to verify whether the technique under investigation is suitable for hyperspectral image fusion. The performance measures from this category can provide some information on the suitability and adaptability of generalized fusion techniques for fusion of a large number of images. These measures include several distance measures between a given fused image and another reference image. We use F_K (ideally, F_k as $k \to \infty$) as the reference image, but denote it by F for the ease and consistency of notations used in this monograph. The evolution of these measures for progressive increase in the number of bands being fused provides a useful information about the consistency of the technique. It should, however, be noted that the consistency analysis in this case refers to the behavior of the fusion technique with respect to a given performance measure only.

The participatory performance measures are evaluated only over the final fused image, and they do not enable us to study the progression of the fusion process. As the name suggests, these measures are used to analyze the degree of *participation* from the bands of input data. A good fusion technique $\mathscr{F}$ is expected to extract useful contents from each of the hyperspectral bands, and combine them appropriately. In this process, the contribution of each of the bands is expected to be similar for fusion process to be robust and accurate. The participatory performance measures indicate the uniformity and symmetry of the fusion technique $\mathscr{F}$ with respect to its constituent images, i.e., whether all the input bands contribute well toward the fusion process, an important aspect often overlooked in the existing literature. A well fused image should have appropriate contributions from all the constituent bands, that is, the fusion technique should be capable of extracting unique information from the bands. This aspect becomes more important with an increasing number of bands, as in hyperspectral images. These measures try to answer questions such as- how the bands are *participating* into fusion? Do all bands participate in fusion? Is the participation uniform across the bands or it is uneven?

In the discussion, we have so far assumed that the hyperspectral data is spectrally ordered in increasing order of wavelengths, which one may refer to as the forward order. The Eq. (9.1) and the subsequent discussions also refer to the incremental fusion in the forward direction. However, the analysis of the technique for the image quality and the consistency also holds true for the incremental fusion of bands in the reverse order. Throughout our discussions, we have not explicitly considered any specific ordering of bands. The spectral ordering is associated with the spatial correlation among successive bands, but does not depend on the forward or reverse ordering. The interpretation and the final result ideally remain exactly the same, and it is largely governed by the spectral signature of a particular pixel. During incremental fusion, the spectral signature of a pixel gets traversed either in forward or reverse direction, but the correlations among the successive bands still exist, which facilitate the application of the performance measures in exactly the same manner. Furthermore, the above argument also holds true for any random permutation of the given set of image bands during fusion. The band-to-band strong correlation no longer persists. In such cases, a large amount of disparate information is contained by the first few randomly permuted bands. Thus, the plots for asymptotic measures tend to saturate quickly as compared to the sequential case.

In the following subsections, we describe each performance measure and its significance in hyperspectral image fusion.

9.3.1 No Reference Quality Measures

The set of no reference measures includes the statistical and other parameters typically used for the evaluation of general image fusion techniques. These measures are evaluated directly over the fused image and no extra information is required. Generally, these measures have a simple physical interpretation which makes them quite popular. For a small number of input images, say 2 or 3, these measures are computed for the resultant image, and the numerical values obtained by applying different techniques are compared. In the case of hyperspectral image fusion, the number of image bands is in the range of hundreds and therefore, it is advisable to study the behavior of these measures over the increasing number of images being fused. Although the relation between the no-reference measures and the quality of the image has been well studied in the literature, here our focus lies in analyzing how these measures reflect the performance of the corresponding fusion technique as the number of constituent input images increases. A brief description of some of these measures and their expected behavior in the context of a set of incrementally fused images is presented below.

1. **Variance:** The variance of an image, $\sigma^2 = var(I)$, is directly related to the image contrast. Variance measures the deviation of gray values of the pixels from the image mean. Images with higher variances have a better contrast, which makes visualization simple and appealing. A smaller variance indicates that gray values of most of the pixels are close to the image mean, and thus, the image mainly consists of a less number of gray value with pixels mostly centered around its mean. The variance σ_k^2 of the k-th incrementally fused image is given by Eq. (9.2).

$$\sigma_k^2 \equiv \sigma^2(F_k) = var\left(\mathscr{F}\left(I_1, I_2, \ldots, I_k\right)\right). \tag{9.2}$$

The variance tends to be higher with addition of noise. Therefore, a high value of σ_k^2 need not necessarily imply a better quality. One needs to be careful while relying completely on the variance measure alone. Notwithstanding above, when calculated over each of the incrementally fused images, the image variance σ_k^2 should ideally increase as the number of constituent images increases.

2. **Entropy:** The entropy of an image, $H(I)$ gives an indication of the information content in the image. Thus, images with a higher entropy are richer in information content. For an image with gray levels $[0, \ L - 1]$, the entropy H_k of the image F_k is given as:

$$H_k \equiv H(F_k) = -\sum_{i=0}^{L-1} \mathscr{P}_{F_k}(i) \ \log \mathscr{P}_{F_k}(i), \qquad (9.3)$$

where $\mathscr{P}_{F_k}$ represents the probability density function of the given image F_k, represented by its gray-level histogram. As more and more images are being fused, we would naturally expect an increase in the information content in the fused image. Thus, an addition of constituent images during the process of fusion should result in an increase in the entropy H_k. Similar to the variance measure, image entropy is also susceptible to the presence of noise in constituent images. Also, the entropy does not provide any information about the spatial details. Its usefulness lies in providing the average information content in the image as a whole.

3. **Average Gradient:** Sharply fused images enhance details in the scene such as edges, boundaries, and various objects. It helps human observers to identify various features, and also improves the performance of various machine vision algorithms such as image segmentation and object recognition. The average gradient of an image $\bar{g}(I)$ is a simple but efficient measure of its sharpness in terms of gradient values. The average gradient is defined [24] by:

$$\bar{g}_k \equiv \bar{g}(F_k) = \frac{1}{X\,Y} \sum_{x=1}^{X} \sum_{y=1}^{Y} \sqrt{\left(\frac{\partial}{\partial x} F_k\right)^2 + \left(\frac{\partial}{\partial y} F_k\right)^2}. \qquad (9.4)$$

The average gradient $\bar{g}_k$ should increase monotonically as k increases. A good fusion technique should ensure that addition of a new band to fusion leads to an increase in sharpness of the fused image. As the isolated noisy pixels tend to increase the spatial gradient, this measure is also susceptible to the presence of noise in the data.

9.3.2 Performance Measures with an Asymptotic Reference

This set of measures is used for the consistency assessment of the fusion technique considering the final fused image $F \equiv F_K$, as the reference. The idea is somewhat different from the conventional ways of fusion evaluation. Also, generalized image fusion typically combines very few images, say 2–3. Not much study has been carried out towards extension of these techniques for larger data sets. We need to analyze the behavior and performance of existing techniques for fusion of large data sets such as hyperspectral images. We proceed as follows. Let us first assume that fusing a large sequence of input images does asymptotically yield an acceptable quality image. The progression of fusion is expected to be smooth and convergent in nature due to strong inter-band correlations in hyperspectral data. During the progression, a few initial image bands impart a higher amount of information towards the resultant. The fusion technique captures most of the information from the initial bands. Thus, the redundancy in the subsequent bands increases as the fusion progresses. The hyperspectral bands, especially towards the end of the fusion process become more redundant, and impart a lesser amount of information toward the resultant. This characteristic of the hyperspectral image data can be easily verified from the analysis of correlation coefficient presented later in this section. The objective of the performance measures presented in this section is to study the progression of the fusion process as more and more image bands are fused. We consider the final fused image (i.e., F_K) as the reference during this study. The objective of these tests is to answer the question: Does the process converge uniformly as desired or does it show any oscillatory behavior? Further, how fast is the progression? For example, a very quick convergence of the sequence implies that the subsequent image bands are unable to provide any additional information during the process. Therefore, a fusion technique providing a very quick convergence is not desirable. These measures determine the suitability of a given technique for the fusion of a large number of bands (or images), but on their own these tests do not provide information about the accuracy of the technique.

1. **Bhattacharyya Distance:** The Bhattacharyya distance is a measure of similarity between two probability density functions (pdf) [84]. This measure computes the extent of overlap between two density functions in order to determine their closeness. In case of digital images, the normalized histogram of the image is considered to be equivalent to the density function of its gray values. The Bhattacharyya distance BD_k for each of the k incrementally fused images with respect to the final image can be calculated. Through this measure, one can quantify the similarity or closeness of the kth incrementally fused image F_k with respect to the final fused image F in terms of their pdfs. As more hyperspectral bands contribute towards the result of fusion, the corresponding incremental fused image is expected to be increasingly similar to the final fused image. As the similarity between the images grows, the overlap between their density functions (or histograms) also increases. For the two discrete probability density functions $\mathscr{P}_k(\zeta)$ and $\mathscr{P}_{\mathrm{ref}}(\zeta)$ over the same domain $\zeta \in \mathscr{Z}$, the Bhattacharyya distance BD_k is defined by Eq. (9.5).

$$BD_k \equiv BD(\mathscr{P}_k, \mathscr{P}_{\mathrm{ref}}) = -\ln\left(\sum_{\zeta \in \mathscr{Z}} \sqrt{\mathscr{P}_k(\zeta)\,\mathscr{P}_{\mathrm{ref}}(\zeta)}\right), \qquad (9.5)$$

where $\mathscr{P}_{\mathrm{ref}}$ is the histogram of the final fused image F, and $\mathscr{P}_k$ is the same for the incrementally fused image F_k. The plot of BD_k as a function of k provides an estimate of the convergence rate of the technique in terms of the pdfs of the images F_k. The value of BD_k reaches the minimum for the final image F, which is obviously zero, and it is maximum at the start. It is expected that the Bhattacharyya distance should monotonically decrease as more image bands are considered indicating a gradually increasing similarity between the images F_k and F.

2. **Jensen-Shannon Distance:** The Bhattacharyya distance requires a reference pdf, and the similarity measure is evaluated with respect to the same. The Jensen-Shannon distance is another method of measuring the similarity between two probability distributions, which eliminates the requirement of the reference distribution [106]. This entity is derived from the Kullback-Leibler divergence that measures the average number of extra bits required to represent the given probability distribution from the reference probability distribution. The Jensen-Shannon divergence, JSD is the symmetricized modification of Kullback-Leibler (KL) divergence. The KL distance of a discrete probability distribution $\mathscr{P}_{Z_2}(\zeta)$ from another distribution $\mathscr{P}_{Z_2}(\zeta)$ defined over the same domain $\mathscr{Z}$ is given by:

$$D_{KL}(Z_2, Z_1) = \sum_{\zeta \in \mathscr{Z}} \mathscr{P}_{Z_2}(\zeta) \log \frac{\mathscr{P}_{Z_2}(\zeta)}{\mathscr{P}_{Z_1}(\zeta)}. \tag{9.6}$$

The Jensen-Shannon distance can be calculated as [106]:

$$JSD(Z_2, Z_1) = \frac{1}{2} D_{KL}(Z_2, Z_{\text{mid}}) + \frac{1}{2} D_{KL}(Z_1, Z_{\text{mid}}), \tag{9.7}$$

where $\mathscr{P}_{Z_{\text{mid}}}(\zeta) = \frac{1}{2}(\mathscr{P}_{Z_1}(\zeta) + \mathscr{P}_{Z_2}(\zeta))$. Similar to the Bhattacharyya distance, the JS distance between each of the k incrementally fused images and the asymptotic reference image can be calculated by Eq. (9.8).

$$JSD_k \equiv JSD(F, F_k) = \frac{1}{2} D_{KL}(F, Z_{\text{mid}}) + \frac{1}{2} D_{KL}(F_k, Z_{\text{mid}}). \tag{9.8}$$

The plot of JSD_k should again monotonically reduce to zero, which indicates that the histograms of the incrementally fused images asymptotically approach the histogram of the final image. The JS distance along with the previous measure (BD) provide the extent of similarity in terms of the histograms (or pdfs) of the images.

3. **Correlation Coefficient:** The correlation coefficient CC has often been used to measure the degree of similarity between two functions [49, 176, 206]. This measure has been used to evaluate the closeness of the fused image to the constituent bands (or images). The performance assessment of the pan-sharpening techniques has also been carried out using the correlation coefficient. We employ this measure CC to study the relation between the incrementally fused images

F_k and the final fused image F_K. The correlation coefficient between each of the k incrementally fused images (F_k) and the final image F indicates the degree of the pixel-level similarity, and thus, it can be used to analyze the consistency of the fusion technique as well. The correlation coefficient has been defined by Eq. (9.9).

$$CC_k \equiv CC(F_k, F) = \frac{\sum_{x=1}^{X} \sum_{y=1}^{Y} (F_k(x, y) - m(F_k)) (F(x, y) - m(F))}{\sqrt{\sum_x \sum_y (F_k(x, y) - m(F_k))^2 \sum_x \sum_y (F(x, y) - m(F))^2}},$$

$$(9.9)$$

where $m(F_k)$ and $m(F)$ are the mean gray levels of images F_k and F, respectively. Similar to the other measures described in this subsection, the similarity between the incrementally fused images and the final fused image is expected to increase as $k \to K$. The coefficient being a normalized measure, the maximum value it can attain is 1, which trivially occurs at the final fused image, i.e., $CC_k|_{k=K} = CC_K = 1$. We expect the function CC_k to approach unity monotonically.

It should be noted that the measures with an asymptotic reference deal with the similarity of the incrementally fused images with the final fused image, and hence the convergence property of the measures. The first two measures indicate the similarity in terms of the image histograms i.e., the gray scale distributions, while the last measure of CC computes the degree of the similarity from the pixel-wise relationship between the two images.

9.3.3 Participatory Performance Measures

Evaluation of fusion quality is an interesting and useful topic. However, the scope of the traditional aspect of quality evaluation is limited to the fused image, and little attention has been paid to how each of the input images (or bands) contribute towards the final result of fusion. This aspect of contribution can be visually judged from the input images and the fused image in case of generalized fusion of a very few images. However, this task is difficult and cumbersome when the number of input images grows to a few hundreds. For efficient fusion, it is necessary that all the input images (or bands) participate towards the fusion process. A good fusion technique should extract the *useful* information content from each of the bands, and combine it to form the resultant image. Understanding the participation from constituent bands is an important aspect of fusion, especially in case of hyperspectral data. If some input bands do not participate towards fusion, one may lose the specific information contained by those bands, and essentially the corresponding details and features will be absent in the final fused image. It is also possible, that such a loss of information might remain unnoticed unless one specifically compares the fused image against all the input bands. Hence, one needs to consider developing certain specific measures that can quantify the relative participation of all constituent bands.

The set of participatory performance measures refers to the quality of the final result of fusion. Therefore, these measures are not useful in studying the progression of the fusion process. Rather, we are interested in the numerical values of these measures only, as opposed to the analysis of plots in case of the previous set of performance measures. The objective of these performance measures is to quantify the contribution or participation of each of the input bands toward building the final image. A good fusion technique is expected to extract suitable information from all the constituent bands during the fusion process. In this section, we explain how this aspect of a fusion technique can be quantified. The set of measures in Sect. 9.3.1 is unable to provide this information. The measures defined in Sect. 9.3.2 provide this information to some extent but do not help in defining the accuracy of the technique.

1. **Relative Bias:** The mean of the fused image $m(F)$, represents the average intensity of the entire image. A higher mean value has been used as a performance indicator in [24, 191]. However, over-saturated images tend to have higher mean values. In such cases, any fusion technique producing a nearly saturated resultant image may be incorrectly considered to be better. Therefore, the mean value is not the correct measure to analyze the performance of the technique. An excessively high value of the mean of the fused image is not required. However, one may expect the mean of the fused image to be close to the mean of entire input data. If we are interested in preserving the radiometric fidelity in the data, we may devise a technique that generates a fused image with mean close to the input mean, and preserve the average radiometric mean. The preservation of mean may not be useful in most of the visual applications, but it does facilitate an easier comparison between different regions across the same and different images. Thus, one should define the deviation of the mean of the resultant image from the mean values of the constituent bands as a performance measure. We suggest the use of the relative bias, which can be calculated as the ratio of sum of deviations in the mean of the constituent bands from the mean of the resultant image to the mean of the resultant image. A lesser deviation in the relative bias indicates closeness of the resultant image to its constituent bands in terms of the average intensity. The relative bias $\hat{b}$, for fusion of hyperspectral data containing K bands can be defined by Eq. (9.10).

$$\hat{b} = \frac{1}{K} \sum_{k=1}^{K} \frac{|m(I_k) - m(F)|}{m(F)}. \tag{9.10}$$

If $m(I_k)$ is the amount of average intensity of the incoming radiation in the k-th spectral band, and $m(F)$ indicates the amount of average intensity of the resultant fused image, then the relative bias $\hat{b}$ measures the radiometric distortion in terms of the mean or average intensity of the bands in the data.

As expected, a simple averaging based fusion yields trivially the least bias and, therefore, this measure alone is not good enough to evaluate the performance of a fusion scheme. It may be noted here that the relative bias as defined here does not explicitly relate to the radiometric fidelity of the fusion process. For such consideration, one may refer to the work by Zhukov et al. which uses the radiometric fidelity of the data as a quality measure while increasing the spatial resolution in the case of pan-sharpening [207].

2. **Fusion Factor:** The fusion factor quantifies the amount of mutual information between each of the input bands (or images) and the fused image. It indicates the extent of similarity of each of the bands with the final fused image, and hence, the contribution of each band towards the final result. The fusion factor FF while fusing two images I_1 and I_2 is defined as [144, 147]:

$$FF(I_1, I_2) = MI(I_1, F) + MI(I_2, F), \qquad (9.11)$$

where the MI is the amount of mutual information between the fused image $F = \mathscr{F}(I_1, I_2)$ using technique $\mathscr{F}$. Chen et al. presented a detailed theoretical analysis of this mutual information-based measure in [37]. A higher fusion factor implies a higher similarity between the final image and the input bands- a necessity for any good fusion technique. However, this definition does not take into account the variation in information contents of individual bands of the hyperspectral data. It is reasonable to expect a smaller value of mutual information for the bands with lesser information content, and vice-versa. To circumvent this problem, we use a modified expression by considering the entropies of the individual bands that indicate the intrinsic amount of information in the corresponding bands [92]. It suggests a weighted addition of the mutual information between the constituent bands and the final image where the weights have been set to the entropy of the corresponding input band. The fusion factor FF is calculated as the normalized weighted sum of the mutual information between input bands and the resultant image. The modified fusion factor assesses the participation of the input bands I_k, $\forall k$ towards the fused image F by appropriately weighing them with respect to their intrinsic information content measured in terms of entropy, i.e., $H(I_k)$. Through a normalizing denominator, it also makes the term independent of the number of constituent image bands. The modified FF for fusion of hyperspectral data set with K bands is given by Eq. (9.12).

$$FF = \frac{\sum_{k=1}^{K} MI(I_k, F)\, H(I_k)}{\sum_{k=1}^{K} H(I_k)}. \qquad (9.12)$$

A higher value of FF, which implies a higher amount of information in F from its constituent bands, is desirable.

3. **Fusion Symmetry:** The measure fusion factor does not provide any idea about the uniformity of the technique in combining the information from all of the constituent images (or hyperspectral bands). A good fusion technique should avoid uneven participation from input bands. In such cases, fused images often highlight features only from a particular subset of hyperspectral bands, while several other features get lost in the process of combining. A higher fusion factor FF does not necessarily indicate a uniform participation or contribution from all

constituent bands. To answer question such as- Do all input images participate well in the process?, a measure called fusion symmetry FS has been defined in [147] for two input images as-

$$FS = \left| \frac{MI(I_1, F)}{MI(I_1, F) + MI(I_2, F)} - 0.50 \right|. \tag{9.13}$$

A lower value of FS is desired, indicating information from both images is symmetrically captured. In order to extend the definition to deal with a large number of images (typically multispectral and hyperspectral), we employ the modified version of the fusion symmetry measure proposed in [92]. The modified fusion symmetry measure quantifies the symmetry in participation of image bands from the entropy weighted standard deviation in the mutual information between the final and the constituent images. Bands with higher information content typically have more features, and hence, are expected to contribute more towards fusion. On the other hand, if a band contains very less features as indicated by a low value of the entropy, we would naturally expect lesser contribution from the same towards the final result of fusion. Therefore, the symmetry in the fusion process must be considered in accordance to the intrinsic information contents of constituent bands. The corresponding definition of fusion symmetry FS in the context of hyperspectral image fusion is given by Eq. (9.14).

$$FS = \sqrt{\frac{\sum_{k=1}^{K} (MI(I_k, F) - FF)^2 \ H(I_k)}{\sum_{k=1}^{K} H(I_k)}}. \tag{9.14}$$

A lower value of this measure is desired, which indicates a uniformity in the fusion process with respect to its constituent images. This definition can be particularly effective in case of remote sensing hyperspectral images where the distribution of information among the bands is usually uneven.

While we have discussed several measures for performance assessment of a fusion technique, it is important to understand their susceptibility to the noise in the data. One is able to quantify the performance of a particular fusion technique with respect to the consistency, sharpness, contrast, etc., using these measures. However, none of these measures can evaluate the robustness of the fusion scheme against noise in the data. If any of the constituent bands is badly corrupted by noise, all measures such as variance, entropy, and average gradient will wrongly indicate a very sharply fused image. A very high value of FF may indicate the presence of possible data corruption in certain image bands, but it is unable to relate the measure to the accuracy of the fusion process.

9.4 Experimental Evaluation

In this section let us study the performances of some of the representative fusion techniques. We illustrate the usefulness of the consistency and performance measures over some of the commonly available fusion techniques and some of the methods discussed in this monograph. Although we have provided some literature survey in Chap. 2, we provide in very brief basic concepts behind some of the other fusion techniques being studied in this section. For more details on the other techniques, readers are requested to refer to Chap. 2, or the corresponding references cited therein.

The simplest technique for displaying the contents of a hyperspectral image is to select three specific image bands from the entire set of image bands in the dataset and assign them to the red, green, and blue channels to form an RGB composite image of the scene. Although this technique does not involve actual fusion of the data, this technique is used to obtain a fast and quick rendering of the data contents (as used by the image browser for AVIRIS). We include this technique for analysis due to its technical and computational simplicity. We shall use the 3-band selection scheme as presented in [49].

We also consider the technique of color matching functions which derives the fusion weights for each of the bands from the hypothesis of how the synthesized image would be seen to the human eye if its range of perceived wavelength were stretched to the wavelength range of the hyperspectral data of interest. The weights here specify the amount of primary colors to be mixed which will create the sensation of the same color as that of viewing the original hyperspectral spectrum [78].

Another technique subjected to performance evaluation in this chapter uses a similar methodology of envelopes as explained for the previous technique of the color matching functions. These envelopes are the piecewise linear functions over the spectral range in order to map deep infrared spectra to magenta hues [79].

The multi-resolution analysis (MRA)–based techniques have proved to be very useful in the field of image fusion. Here, we consider an MRA-based technique where each of the constituent image bands is decomposed to extract directional gradient information at each level [189].

We also analyze the edge-preserving technique for fusion of hyperspectral image described in Chap. 3. This technique uses an edge-preserving bilateral filter for the calculation of weights for each of the pixels in each of the image bands [88]. This bilateral filtering-based technique has been discussed in Chap. 3 of the monograph, which happens to be the computationally fastest among all the techniques discussed in this monograph. The choice of the above set of fusion techniques is based on the fact that these are some of the recent techniques for pixel-level fusion of hyperspectral data and that together they span almost all different methodologies of image fusion, such as subspace-based, MRA-based, and local contrast-based techniques; along with the representative technique of band selection for display. While some of these techniques are capable of producing RGB outputs, we deal with the grayscale results of fusion in order to maintain uniformity in the evaluation.

The present chapter primarily focuses on the consistency analysis of the fusion techniques, and illustrates the same over only two hyperspectral datasets for the purpose of brevity. We use the moffett$_2$ dataset from the AVIRIS hyperspectral imager, and the geological dataset from the Hyperion imager. The detailed description of these two datasets can be found in Chap. 10.

Consider the plot of variance as more and more bands are sequentially added towards fusion of the moffett$_2$ data for various techniques as shown in Fig. 9.1a. As expected, the variance increases as the information from more numbers of image bands is being combined. Figures 9.1b, c depict the plots for entropy and for average gradient, respectively for the same dataset, moffett$_2$. One can observe a similar nature of these performance measures from these plots. A larger fluctuation in these plots for the incremental fusion violates the consistency property defined in Sect. 9.3.2. Therefore, one would ideally expect higher values of these no-reference performance measures, but with lesser and smaller fluctuations. Also, we can observe a nearly monotonically increasing nature of these plots, which confirm the consistency of these techniques for the given performance parameter. However, it may be noted that the technique based on the selection of three bands is not consistent as there are large variations in the corresponding plots. As more bands are added, it selects totally different subsets of bands for visualization, and these bands are not well correlated to the results at the previous step. As a matter of fact, one may observe a similar behavior for PCA-based techniques, signifying that the subspace-based techniques are not consistent. We do not discuss PCA-based techniques, as they turn out to be computationally very demanding. Further, Fig. 9.1c suggests that average gradient being a derivative-based measure is quite susceptible to the presence of noise in data, and is not a good performance measure. The measure tends to saturate too quickly and, hence, is not discriminatory enough.

Figure 9.2a shows the variation in the Bhattacharyya distance for the incrementally fused images of the moffett$_2$ dataset. As explained earlier, the final result obtained through fusion of the entire data using the same fusion technique has been considered as the reference for the corresponding technique. The plots of the Bhattacharyya distance for different techniques show a gradual decrease in the value of the measure as more images are used. Asymptotically, the value of Bhattacharyya distance reaches zero. It can be observed that most of the plots of the Bhattacharyya distance have fluctuations till initial 40–50 bands get fused. Thus, the initial bands have larger contributions towards the final fused image. The contribution from subsequent bands gradually reduces, and the corresponding incrementally fused images do not change much as k increases. An analysis of the Jensen-Shannon distances [Fig. 9.2b] of these images from the final output reveals a similar insight into the convergence rate of the process. As expected, the JS distance asymptotically approaches zero. The asymptotic behavior of the measure of correlation coefficient between the set of corresponding incrementally fused images and the final image can be seen in Fig. 9.2c. The fairly non-decreasing nature of these plots again confirms, prima facie, the suitability of these techniques, except the one based on band selection, for the hyperspectral data.

Fig. 9.1 Quality assessment for various techniques for fusion of hyperspectral data with no reference measures for the moffett$_2$ dataset from the AVIRIS. The legends are as follows: Blue (continuous) lines represent the technique of three-band selection. Green lines represent the bilateral filtering technique. Color matching function technique is represented by dashed red lines, while dashed brown lines are indicators of the MRA-based technique. Black (continuous) lines represent the piecewise linear function technique. (Color is viewable in e-book only.) **a** Variance. **b** Entropy. **c** Average gradient (©2011 Elsevier, Ref: [92])

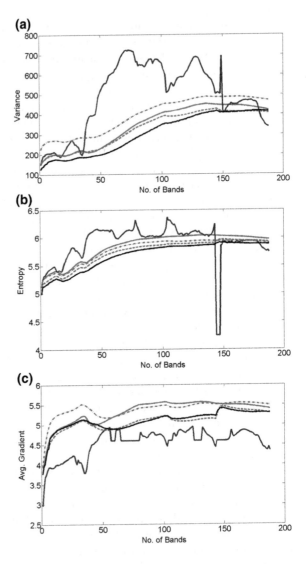

Figures 9.3 and 9.4 represent the plots of the performance measures for the geological image from the Hyperion. The nature of the performances for variance and entropy measures can be observed from Figs. 9.3a, b. This monotonically increasing nature of these no-reference measures is similar to the observations for the moffett$_2$ dataset. These plots highlight the uncorrelated nature of selected bands for synthesis using the band selection technique. Further, it may be noticed that for both test datasets, the performance of the MRA-based technique changes a bit unpredictably at the beginning and then recovers as more bands are used. When the number of bands is relatively lower, the changes in local weights used to combine the high frequency

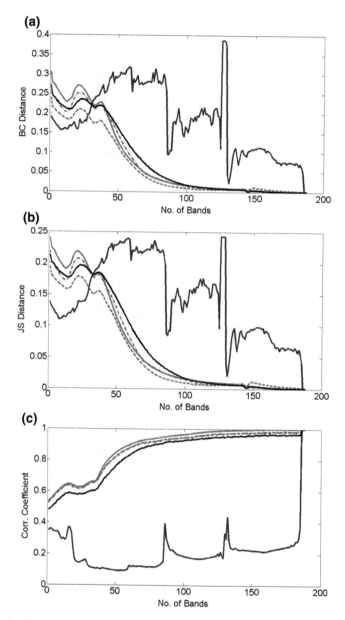

Fig. 9.2 Quality assessment for various techniques for fusion of hyperspectral data with asymptotic reference measures for the moffett$_2$ dataset from the AVIRIS. The legends are as follows: Blue (continuous) lines represent the technique of three-band selection. Green lines represent the bilateral filtering technique. Color matching function technique is represented by dashed red lines, while dashed brown lines are indicators of the MRA-based technique. Black (continuous) lines represent the piecewise linear function technique. **a** Bhattacharyya distance. **b** Jensen-Shannnon distance. **c** Correlation coefficient (©2011 Elsevier, Ref: [92])

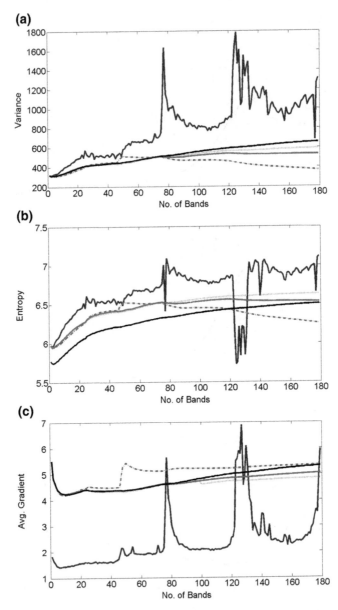

Fig. 9.3 Quality assessment for various techniques for fusion of hyperspectral data with no reference measures for the geological dataset from the Hyperion. The legends are as follows: Blue (continuous) lines represent the technique of three-band selection. Green lines represent the bilateral filtering technique. Color matching function technique is represented by dashed red lines, while dashed brown lines are indicators of the MRA-based technique. Black (continuous) lines represent the piecewise linear function technique. **a** Variance. **b** Entropy. **c** Average gradient (©2011 Elsevier, Ref: [92])

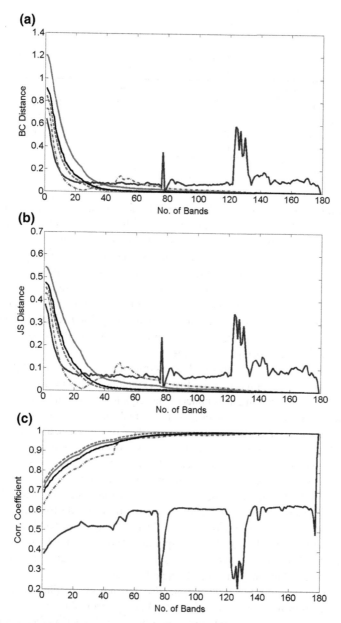

Fig. 9.4 Quality assessment for various techniques for fusion of hyperspectral data with asymptotic reference measures for the geological dataset from the Hyperion. The legends are as follows: Blue (continuous) lines represent the technique of three-band selection. Green lines represent the bilateral filtering technique. Color matching function technique is represented by dashed red lines, while dashed brown lines are indicators of the MRA-based technique. Black (continuous) lines represent the piecewise linear function technique. **a** Bhattacharyya distance. **b** Jensen-Shannnon distance. **c** Correlation coefficient (©2011 Elsevier, Ref: [92])

Table 9.1 Performance measures for various techniques for visualization of the moffett$_2$ data (©2011 Elsevier)

Fusion technique	Relative bias $\hat{b}$	Fusion factor FF	Fusion symmetry FS
Bilateral filtering technique	0.22	1.62	0.24
Three band selection	0.46	1.01	0.66
Piecewise linear function	0.20	1.64	0.29
Color matching function	0.22	1.64	0.26
MRA technique	0.24	1.18	0.18

details is quite significant for this dataset and this affects the performance. However, as the number of bands progressively increases, the averaging effect helps reduce the fluctuations. The spatial domain-based techniques do not suffer much from this problem. Figure 9.4 focuses on the consistency analysis of the fusion techniques. The plots of the Bhattacharyya distance and the Jensen-Shannon distance from Fig. 9.4a, b, respectively, appear monotonically decreasing, finally converging to zero. These plots, thus, affirm the consistency of the techniques (except the band selection technique), and also indicate their suitability for fusion of a very large number of images. Thus, the techniques proposed in [79, 88] prove to be quite suitable for the fusion of hyperspectral images.

The plots for various measures with an asymptotic reference do not provide any numerical values for comparison of fusion techniques. However, these plots are quite useful in understanding the impact of subsequent image bands in the process of fusion. Ideally, we want the fusion process to gradually converge over the entire set of hyperspectral bands. In this case, the corresponding plot has a nearly uniform slope. A plot with very high slope on the either end points out negligible contribution of image bands on the corresponding other end towards the final result. While we have discussed the ordered hyperspectral data where the contents of the bands change gradually, an interesting case would be to observe the effect of fusing the bands in a random order. Through the random permutation, we destroy the inter-band correlation characteristic of the hyperspectral data. If the image bands are randomly permuted and then combined, the corresponding plots will tend to saturate quite quickly as the subsequent bands are no longer highly correlated and the process starts seeing the redundancy in bands being added later.

The performance of a fusion technique also depends on the dataset. This can be illustrated easily from these two test datasets. A comparison of Figs. 9.1–9.4 suggests that all these plots tend to saturate quite rapidly for the Hyperion data compared to the AVIRIS data. In the case of geological Hyperion data, one recovers very little additional information after fusing around 50–60 bands The MRA and the piecewise linear function–based techniques were found to saturate at a slower rate compared to the bilateral filtering-based technique from Figs. 9.1 and 9.2. For the AVIRIS data, the process does not saturate even after fusing 120–130 image bands. This brings out the data dependencies of the performance of any fusion technique.

Table 9.2 Performance measures for various techniques for visualization of the geological data
(©2011 Elsevier)

Fusion technique	Relative bias $\hat{b}$	Fusion factor FF	Fusion symmetry FS
Bilateral filtering technique	0.21	1.59	0.19
Three band selection	0.10	1.33	0.44
Piecewise linear function	0.24	1.71	0.38
Color matching function	0.23	1.75	0.35
MRA technique	0.18	1.36	0.16

For a comparison of the numerical values, we provide the results of the participatory performance measures for the final images F, using all these studied techniques for the Moffett data Table 9.1, and the Geological data Table 9.2. The multi-resolution analysis (MRA)-based technique turns out to be superior to most of the other techniques in terms of relative bias $\hat{b}$ and fusion symmetry FS. However, a very low value of the fusion factor FF indicates a smaller amount of mutual information between the final image and the constituent bands. Therefore, although the participation from the image bands is quite uniform, the quantitative gain from the bands in terms of information is less as indicated by low values of the fusion factor.

The bilateral filtering-based technique performs well in terms of low values of the relative bias and fusion symmetry, indicating that the final image has a lower derivation from the constituent images in terms of intensities, as well as information content. High values of fusion factor FF indicate a significant amount of mutual information between the constituent bands and the fused image.

The techniques using piecewise linear functions and color matching functions provide results with a high variance yielding a good contrast in the final results (will be discussed in the next chapter). The objective assessment shows that both techniques offer a high value of fusion factor (FF), but values of the fusion symmetry FS are also large, which is not desirable. This indicates an uneven participation from input hyperspectral bands. Thus, not all the bands contribute uniformly toward the fusion process. Additionally, the values of the fusion symmetry parameter for these two techniques were found to vary over the test datasets as well. The technique of piecewise linear function and the color matching function assign a constant fusion weight to each of the constituent bands. In case of the piecewise linear function technique, these fixed weights are derived from the optimization of certain criterion in the sRGB space, while the technique of color matching function assigns these constant weights to the particular band by modeling certain perceptual properties of the human eye. In both cases, however, the calculation of weights is fixed and does not consider the local variation in data contents at all. Quite naturally, the subset of bands receiving higher weights has a higher contribution towards the fused image. Therefore, features from this subset of bands find a good representation in the fused image. On the other hand, features in rest of the bands are likely to contribute very little as the corresponding bands have smaller fusion weights. Therefore, one can obtain very good fusion results with such techniques where the subset of bands getting a higher weightage coincides with the subset of hyperspectral data containing high amounts of information. However, in other cases, these techniques may unnecessarily highlight a subset of bands does not contain enough useful features. In contrast to this, the bilateral filtering-based technique and the MRA-based technique evaluate the local neighborhood of each and every pixel in the data for the calculation of fusion weights. Hence, the later techniques prove to be more consistent with the quality of the result produced.

For visual illustration of the *consistency* of a technique, we provide some of the incrementally fused images of the moffett$_2$ dataset using the bilateral filtering-based technique. The result of fusion of first 50 bands of the moffett$_2$ dataset can be seen in Fig. 9.5a. The resultant image from fusion of 100 bands of the same data has been provided in Fig. 9.5b. A comparison between these two images gives an idea about the contribution of subsequent bands. A fusion of first 50 bands provides the outline of the rough shape of the objects present in the data, while subtle details and finer boundaries appear clear in Fig. 9.5b where 100 bands have been combined. This effect is more clear in the top left and right corners of the image which contain a

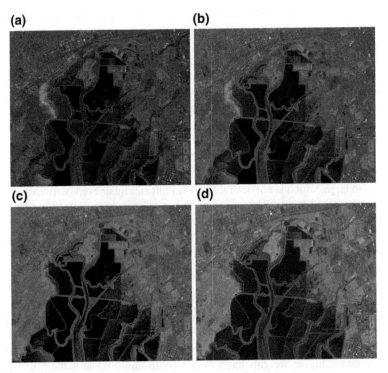

Fig. 9.5 Results of bilateral filtering-based incremental visualization of the moffett$_2$ data. **a, b** and **c** show the results of fusion of 50, 100, and 150 bands, respectively. **d** shows the final fused image (©2011 Elsevier, Ref: [92])

large number of small rectangular objects, which were almost not seen in Fig. 9.5a. A resultant image from the fusion of 150 bands is shown in Fig. 9.5c, which provides a still sharper and clearer understanding of the scene contents. It can also be observed that the subsequent image generated from fusion of the entire dataset as shown in Fig. 9.5d is not too different from Fig. 9.5c. This indicates saturation in the visualization process for the moffett$_2$ dataset for a given technique. The corresponding plots of the asymptotic reference measures given in Fig. 9.2 are also in agreement with the process of saturation.

It is hoped that the results of these and similar experiments can assist an observer in quantifying the efficacy of these quality measures, and in analyzing the performance of various fusion techniques. The discussed quality measures help in capturing the complementary characteristics of a given technique and, thus, instead of relying on the use of a single quality index, one should use a subset of these measures for the selection of the most suitable technique.

9.5 Summary

This chapter defines the notion of consistency of a fusion technique that refers to the asymptotic behavior of the technique with respect to a specific performance measure. The analysis of the consistency can be useful to test the effectiveness and applicability of the given technique for the fusion of a large number of images. Several measures related to the consistency can be used to determine its suitability for fusion of hyperspectral data. For the fusion techniques studied in this chapter, we found that the band selection-based methods, including those of PCA-based methods, are not consistent. However, the other methods such as the MRA-based and the bilateral filtering-based methods are consistent.

We have also presented a detailed analysis of several performance measures as regards the fusion technique. Certain modifications in some of these measures have been discussed for a better analysis of the fusion technique, particularly extending their applicability to the fusion of hyperspectral images where a very large number of image bands are fused. Again, among the techniques analyzed here, the band selection-based techniques appear to perform poorly in terms of the measures of the resultant image compared to the MRA- and bilateral filtering-based techniques.

Chapter 10
Results and Discussions

10.1 Introduction

This chapter provides the consolidated results of various techniques for visualization-oriented fusion of hyperspectral images presented in Chaps. 3, 5–7 of this monograph. These fusion techniques together span a wide variety of image processing methodologies. We began with a signal processing approach, and derived the α-mattes for fusion using a bilateral filter. The second technique dealt with fusion as an estimation problem, and provided a solution using the Bayesian framework. We also explored the concept of matte-less fusion in Chap. 6 where there was no explicit calculation of fusion weights. Lastly, we posed fusion as an optimization problem where some of the desired characteristics of the fused output image had driven the process of fusion. Combining the results from all these solutions enables us to compare and analyze their performances. We also consider some of the other recently developed techniques of hyperspectral image fusion for comparison along with the techniques discussed in this monograph. The visualization technique related to the selection of three specific bands chosen according to a certain set of criteria is quite popular for rendering the contents of the hyperspectral scene. Though such techniques do not involve any kind of fusion, and have been shown to be inconsistent, these techniques are computationally simpler and easy to operate. We present the results of the three band selection technique described in [49] along with the other results for the performance assessment. Two very specific techniques for displaying the hyperspectral image have been proposed in [78, 79]. These techniques generate the fused image from a linear combination of the spectral bands. In the first technique, the fusion weights have been calculated from a set of piecewise linear functions which are based on certain optimization criteria for display on the standard color space (sRGB), and the perceptual color space (CIELAB). The second technique developed by the same authors defines the fusion weights from the CIE 1964 tristimulus functions by stretching them across the wavelength axis in order to cover the entire range of the hyperspectral image. We provide the results of fusion of hyperspectral images using these two techniques for comparisons. The quantitative performance indices

S. Chaudhuri and K. Kotwal, *Hyperspectral Image Fusion*,
DOI: 10.1007/978-1-4614-7470-8_10, © Springer Science+Business Media New York 2013

explained in Chap. 9 bring uniformity to such comparisons. This chapter presents a comparative study using the same set of performance measures.

Each of the fusion solutions requires a certain set of parameters which control the quality of the fused image. We have used the same parameter values for the implementation of each of the fusion solutions described in respective chapters of this monograph. For other techniques, we have used the same parameters as described in the corresponding references cited therein. While it is possible to obtain either grayscale or RGB results of the fusion using any of the fusion techniques from the monograph, most of the performance measures have been defined for the grayscale images. Thus, we discuss the quantitative assessment of the resultant grayscale images of all the techniques considered in this chapter. The RGB images, however, facilitate an easy and better visual comparison. Therefore we also illustrate the color results of the fusion techniques discussed here. The readers are requested to go through the e-book version of the monograph for an RGB pseudo-color representation of the fused images. Also, the appearance of the color images varies with the settings of display and printing devices. The readers are requested to take a note of this.

10.2 Description of Hyperspectral Datasets

For any of the fusion solutions discussed in this book, we do not assume any prior knowledge about the scene or the imaging system. The original hyperspectral image contains nearly 200+ bands. However, some of these bands have a near zero response as the molecular absorption bands of water and carbon dioxide block the transmission of radiation in certain wavelength bands [160]. In addition to those, data in some bands are highly corrupted which appear like noise. After removal of these bands, the number of available bands reduces to around 175–180. Additionally, we normalize the hyperspectral data, i.e., the dynamic range of the pixel has been scaled to unity through an appropriate scaling of the whole dataset.

In this chapter, we provide extensive experimental results for 6 datasets from two different hyperspectral devices. The first hyperspectral sensor is the Airborne Visible/Infrared Imaging Spectrometer (AVIRIS) operated by the National Aeronautics and Space Administration/Jet Propulsion Laboratory (NASA/JPL).[1] The datasets provided by the AVIRIS contain 224 bands where the size of each band is (614×512) pixels. The first two datasets by the AVIRIS depict some regions of the Moffett Field, CA. We refer to these data as the $moffett_2$, and the $moffett_3$ data, respectively. The $moffett_2$ data have already been discussed in the last chapter for illustrations of the consistency of fusion techniques. Another dataset provided by the AVIRIS captures the area of the Lunar lake, NV. These data which depict underwater features have been referred to as the $lunar_2$ in this chapter.

---

[1] AVIRIS data available from JPL/NASA: http://aviris.jpl.nasa.gov.

Table 10.1 Description of the hyperspectral datasets used in this monograph

Data name	Sensor	Dimensions	Geographical location	Region depicted
Moffett$_2$	AVIRIS	$614 \times 512 \times 224$	37°44′N 121°80′W	Moffett Field, CA
Moffett$_3$	AVIRIS	$614 \times 512 \times 224$	37°44′N 122°21′W	Moffett Field, CA
Lunar$_2$	AVIRIS	$614 \times 512 \times 224$	38°44′N 115°78′W	Lunar Lake, NV
Geological	Hyperion	$512 \times 256 \times 242$	37°47′N 117°49′W	West-central Nevada,
Coral	Hyperion	$512 \times 256 \times 242$	20°24′S 164°30′E	Coral Sea, New Caledonia
Urban	Hyperion	$512 \times 256 \times 242$	37°47′N 122°13′W	Palo Alto, CA

We choose the remaining 3 datasets from the Hyperion imaging sensor used in the EO-1 spacecraft for earth observation.[2] These data consist of 242 bands covering the bandwidth from 0.4 to 2.5 μm. The data processed for the terrain correction are designated as the Level G1. The dimensions of bands in the original are (2905×256) pixels. However, for faster computation, we split the data across their length to obtain several datasets of a smaller dimensions (512×256) pixels. We select datasets depicting a wide variety of features. The first data provide some of the geological features of west-central Nevada area, and hence it will be called as the geological data. We have already presented this dataset for the consistency analysis of a fusion scheme in Chap. 9. We also select a hyperspectral dataset capturing some regions of the coral reefs near Australia. This dataset has been referred to as the coral data. Since we have discussed four different techniques for fusion of hyperspectral data in Chaps. 3, 5–7 of this monograph, we illustrate the corresponding results of fusion finally using the urban dataset for every solution. This urban dataset is obtained by the Hyperion, which captures some of the urban regions of Palo Alto, CA, and hence the name. Details of these datasets can be found in Table 10.1.

10.3 Results of Fusion

Let us begin with the performance evaluation of the moffett$_2$ data. The results of fusion using the techniques described in previous chapters are shown in Fig. 10.1. The bilateral filtering-based technique provides a sharp image of the scene as shown in Fig. 10.1a. The moffett$_2$ scene contains a large number of smaller objects mainly on the top and the right sides of the image, which are well-discriminable in this image. The output of the Bayesian fusion technique from Fig. 10.1b, however, appears sharper than the first image. Although this image has a slightly brighter appearance, the sharpness in the image appears to be independent of the higher mean value. This sharpness is due to the fact that the pixels with higher gradient values have contributed more towards the resultant final image. The appearance of the output of

---

[2] Hyperion data available from the U.S. Geological Survey:http://eo1.usgs.gov.

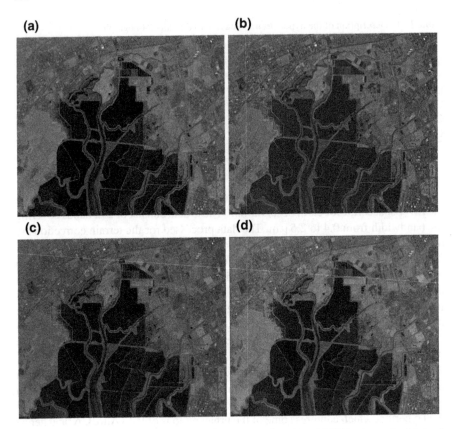

Fig. 10.1 Results of visualization of the moffett$_2$ data from the AVIRIS using **a** the bilateral filtering-based technique, **b** the Bayesian technique (reproduced from Fig. 5.3b), **c** the variational technique, **d** the optimization-based technique (reproduced from Fig. 7.2b), **e** the three band selection technique, **f** the piecewise linear function technique, and **g** the color matching function technique. The gamma is set to 1.50 for all color displays (Color is viewable in e-book only)

the variational fusion technique shown in Fig. 10.1c is similar to that of the Bayesian fusion in terms of contrast and clarity in the scene where the smaller objects in the scene are well identifiable. The fused image using the optimization-based solution is shown in Fig. 10.1d. The high amount of visual contrast in this image is due to the specific objective function in the problem formulation. This image appears somewhat different in terms of its color composition. However, the coloring scheme being a pseudo one, it does not reflect the quality of fusion. It may be observed that the band selection technique essentially captures a very small fraction of data. Therefore, the corresponding resultant image from Fig. 10.1e does not contain enough features, and thus, lacks clarity. This fact will be more evident when this image is compared with the other results. Figures 10.1f, g provides the results of fusion of the moffett$_2$ data using the piecewise linear functions (PLF), and the color matching functions (CMF),

(e)   **(f)**

(g)

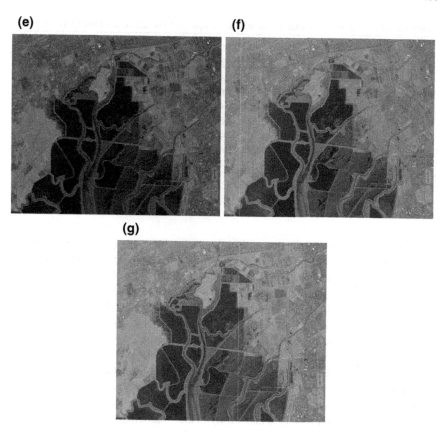

Fig. 10.1 (continued)

respectively. As these techniques operate over all the bands, most of the data features have been captured for fusion. However, we observe a high amount of saturation at some places in both results.

The performance assessment of these results using the no-reference measures and the participatory measures described in Chap. 9 can be found in Table 10.2. We have already discussed the participatory performance measures for the moffett$_2$ dataset in Chap. 9. Here we discuss in brief only the no-reference measures. The values of participatory measures have also been provided for a quick reference. All the four presented solutions provide high values of variance σ^2 which indicate a good amount of contrast in them. Comparatively low values of the variance for the techniques of PLF and CMF are due to over-saturation in these results. The moderate values of the average gradient $\bar{g}$ for these techniques can also be explained with a similar reasoning. These two measures quantify the amount of contrast and the sharpness in the images, respectively. The entropy values H are quite high for the results of all techniques, however the value is the lowest for the result of the three band selection technique.

Table 10.2 Performance measures for various techniques for visualization of the moffett$_2$ data

Fusion technique	Variance σ^2	Entropy H	Avg gradient $\bar{g}$	Relative bias $\hat{b}$	Fusion factor FF	Fusion symmetry FS
Bilateral filtering technique	476.86	6.02	4.65	0.22	1.62	**0.24**
Bayesian technique	487.65	5.68	**6.10**	0.33	**1.81**	0.30
Variational technique	467.94	5.95	4.75	**0.17**	1.79	0.32
Optimization technique	**593.04**	**6.12**	5.52	0.39	1.62	0.32
Three band selection	481.41	5.59	4.86	0.45	1.01	0.66
Piecewise linear function	421.35	5.93	4.35	0.20	1.64	0.29
Color matching function	403.44	5.94	4.34	0.22	1.64	0.26

(Bold font denotes the best performance)

It can be inferred that the selection of three bands for display may lack in information content, but it may not bear any relationship with the sharpness or the contrast in the image. It may select the bands possessing high amounts of visual contrast and sharpness. The values of relative bias $\hat{b}$ are low for all the techniques except the band selection and the optimization-based one. The selection of bands deals with only a small fraction of the input hyperspectral data, and thus, it does not perform well in terms of the participatory performance measures. The optimization-based solution explicitly deviates from the relative bias and the fusion symmetry (FS) in order to generate an output that is intended to maximize certain output characteristics. The variational technique too, iteratively modifies the fused image for well-exposedness. Though this indicates a good participation from the constituent input bands (in the form of a higher fusion factor), a uniformity in their participation is not guaranteed, as indicated by a comparatively higher value of fusion symmetry. Therefore, the values of participatory measures in the case of these two techniques do not match well with the qualitative description seen in Fig. 10.1.

It may also be noted that the fusion factor (FF) that reflects the amount of mutual information between the fused image and the input spectral bands is quite high for all techniques except the band selection technique.

We now provide the results of fusion over the second dataset—the moffett$_3$ data in Fig. 10.2. These data depict the region in surroundings of the first dataset (see the latitude–longitude information in Table 10.1). Therefore, the variety in the features, and thus, the nature of corresponding fusion results are quite similar to those of the moffett$_2$ data. The only notable difference lies in the colors of the results of the band selection technique and the CMF technique. We restate that the coloring schemes employed for all the results are purely pseudo-coloring schemes, and hence the change in the image due to the color or hue change should not be construed as a measure of its quality. However, the change of colors in the result of band selection given in Fig. 10.2e is due to different choice of bands as compared to the ones in case of the moffett$_2$ data. The saturation in the results of the technique using the PLF is still persistent in Fig. 10.2f. While the results of the bilateral filtering-based fusion shown in Fig. 10.2a are visually appealing, some regions in the right bottom corner seem to lack in contrast. This area appears well-contrasted and thus, is quite clear in the results of the other two solutions as shown in Fig. 10.2b–d. These techniques explicitly consider the contrast measure either from the input or from the output perspective, which leads to such improvements over the other techniques. Figure 10.2c shows the result of the variational fusion technique. Although it is similar to the result of the bilateral filtering-based technique, it lacks in sharpness, and thus, smaller objects in the right side of the scene are not very clearly observable.

Higher values of variance σ^2, and average gradient $\bar{g}$ for the Bayesian and the optimization-based techniques seen in Table 10.3 are in agreement with the corresponding results in Fig. 10.2. However, we observe similar values of entropy for all the results, which indicate a similar amount of average information content for all images. The fusion factors (*FF*) for all the presented techniques indicate their usefulness towards producing results with higher contents from the input data.

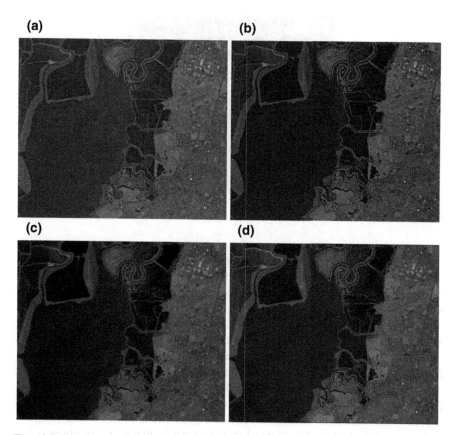

(a) **(b)**

(c) **(d)**

Fig. 10.2 Results of visualization of the moffett$_3$ data from the AVIRIS using **a** the bilateral filtering-based technique, **b** the Bayesian technique (©2013 Elsevier, Ref. [91]), **c** the variational technique, **d** the optimization-based technique, **e** the three band selection technique, **f** the piecewise linear function technique, and **g** the color matching function technique

(e)   **(f)**

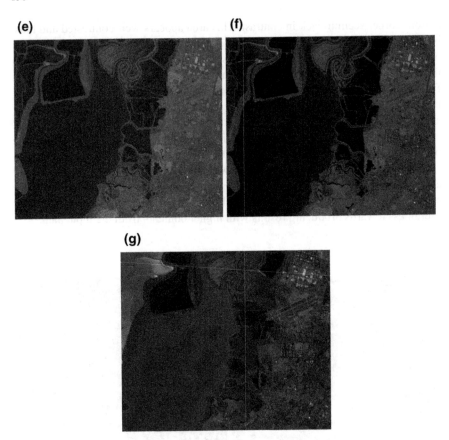

(g)

Fig. 10.2 (continued)

The relative bias $\hat{b}$ and the fusion symmetry (*FS*) for the band selection and the optimization-based techniques are quite high for the same reasons explained in the case of the moffett$_2$ data. The bilateral filtering-based technique and the variational technique do not deviate from the mean of the input hyperspectral image as opposed to the other two techniques where this deviation is explicitly imposed for an efficient visualization. The variational technique provides the least deviation in terms of the relative bias $\hat{b}$ as the technique has been designed to maintain the overall saturation in the fused image close to that of the input hyperspectral bands.

Let us now consider the lunar$_2$ dataset from the AVIRIS. This dataset is different from the previous datasets of the Moffett Field in terms of scene contents. This dataset depicts some underwater features of the Lunar lake. The results of fusion using the techniques discussed in this monograph are shown in Fig. 10.3. We observe that almost all these results are visually similar. The lunar data do not contain much features as opposed to the first two datasets. Additionally, most of these features are present over a substantially large number of spectral bands. Therefore, the composite

Table 10.3 Performance measures for various techniques for visualization of the moffett$_3$ data

Fusion technique	Variance σ^2	Entropy H	Avg gradient $\bar{g}$	Relative bias $\hat{b}$	Fusion factor FF	Fusion symmetry FS
Bilateral filtering technique	374.29	5.60	4.62	0.31	1.75	**0.21**
Bayesian technique	451.84	5.17	**4.82**	0.32	1.93	0.94
Variational technique	385.35	5.63	3.53	**0.18**	1.95	0.26
Optimization technique	**527.12**	**5.71**	4.47	0.40	**2.03**	0.98
Three band selection	363.64	5.24	3.30	0.36	1.29	0.49
Piecewise linear function	282.05	5.55	3.18	0.45	1.47	0.47
Color matching function	157.50	5.19	2.59	0.44	1.24	0.28

Table 10.4 Performance measures for various techniques for visualization of the lunar$_2$ data

Fusion technique	Variance σ^2	Entropy H	Avg gradient $\bar{g}$	Relative bias $\hat{b}$	Fusion factor FF	Fusion symmetry FS
Bilateral filtering technique	642.93	6.32	4.59	0.24	1.65	0.08
Bayesian technique	630.91	6.47	4.43	0.19	1.42	**0.04**
Variational technique	611.25	6.41	4.22	**0.14**	2.12	0.36
Optimization technique	**669.91**	6.40	**5.02**	0.23	2.14	0.15
Three band selection	658.49	6.47	4.35	0.21	1.83	0.11
Piecewise linear function	642.35	6.45	4.61	0.16	1.70	**0.04**
Color matching function	612.49	**6.56**	4.47	0.16	**2.15**	0.06

formed by the specifically selected three bands as shown in Fig. 10.3e is capable of capturing most of the data contents. The bilateral filtering-based solution particularly enhances the weaker features in the data, and therefore, provides somewhat higher amount of visual information than the other techniques. These minor differences in the quality of the fusion results are observable from the quantitative performance measures provided in Table 10.4. All the techniques generate fused images with similar values of variance and entropies. The fusion factor FF and fusion symmetry FS of these techniques indicate comparable participation of bands for all the techniques. The result of variational technique possesses the smallest value of the relative bias by virtue of the iterative correction in the mean of the fused image, which in turn measures the deviation of the output intensity from that of input bands. Higher value of the fusion factor for this technique also indicates a large amount of information transfer from the input bands towards the fusion output.

The optimization-based technique yields the highest value of the average gradient than any other technique, which results in a better perceived sharpness in Fig. 10.3d. This technique processes the hyperspectral data for obtaining an image with high contrast, irrespective of the input characteristics. Therefore, we are able to obtain a high value of the variance in the result despite lack of contrast within the input bands. As we may note, this advantage comes at the cost of a higher value of fusion symmetry.

Having discussed the results of processing the AVIRIS data, we consider the remaining three hyperspectral datasets provided by the Hyperion imaging sensor.

(a)   **(b)**

(c)   **(d)**

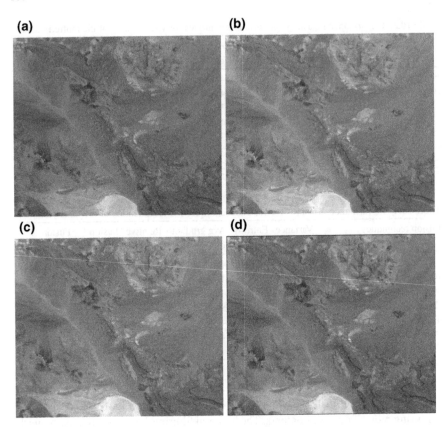

Fig. 10.3 Results of visualization of the lunar$_2$ data from the AVIRIS using **a** the bilateral filtering-based technique, **b** the Bayesian technique, **c** the variational technique, **d** the optimization-based technique (©2012 IEEE, Ref. [90]), **e** the three band selection technique, **f** the piecewise linear function technique, and **g** the color matching function technique

We begin with the results of fusion over the geological dataset. The geological data contain less number of features, similar to that of the lunar$_2$ hyperspectral image. For such datasets, obtaining an image with high contrast and higher details is quite a challenging task. The fusion technique should be able to selectively extract the necessary details from the data, and appropriately represent them in the resultant image. The results of fusion using various techniques can be seen in Fig. 10.4. The bilateral filtering-based technique identifies the fine textural contents in every band, and uses them to define the fusion weights. Therefore, the corresponding result can be seen to have preserved a large number of finer details in Fig. 10.4a. The result of the Bayesian method as seen in Fig. 10.4b appears to be quite appealing through visual examination. The gray appearance of this image indicates similarity in contents of the three images each fused from one-third data to form the RGB output. The variational technique also brings out various minor details in the scene as shown in Fig 10.4c.

(e) **(f)**

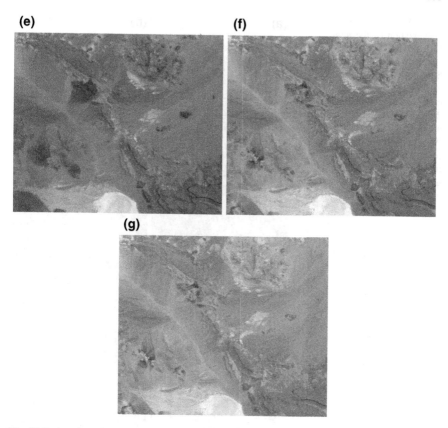

(g)

Fig. 10.3 (continued)

However, as the technique attempts to preserve the overall intensity of the fused image with reference to the input data, the result appears a bit dull. The output of the optimization-based technique, too, provides a high amount of contrast, and proves to be better for visual interpretation. The results of the PLF and the CMF techniques as shown in Fig. 10.4f–g, respectively, have also captured most of the details from the data. However, these two results lack in contrast and sharpness, giving the images a washed-out appearance.

An objective assessment of these results can be found in Table 10.5. The numerical values of the no-reference measures are in agreement with our visual observations. All the presented techniques yield fused images with high values of variance, entropy, and sharpness indicating their usefulness in terms of contrast, average information, and the perceived sharpness, respectively. The highest values of the gradient ($\bar{g}$), and the variance (σ^2) are observed for the optimization-based technique which reflects its ability to provide visually appealing results even from the data containing less features. However, none of the techniques guarantee nearly equal contributions from the constituent hyperspectral bands as the values of fusion symmetry (*FS*) are quite

Fig. 10.4 Results of
visualization of the geo-
logical data from the Hyperion using **a** the bilateral
filtering-based technique,
b the Bayesian technique,
c the variational technique,
d the optimization-based technique (©2012 IEEE, Ref.
[90]), **e** the three band selection technique, **f** the piecewise
linear function technique, and
g the color matching function
technique

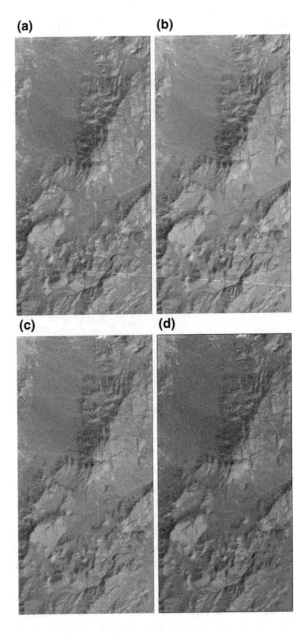

high except for the bilateral filtering-based technique. The values of the fusion factor
(*FF*) which reflects the amount of information transferred from the input to the output
are quite high for the Bayesian, variational, and the optimization-based solutions.
This indicates that although not all the bands contribute equally towards fusion for

Fig. 10.4   (continued)

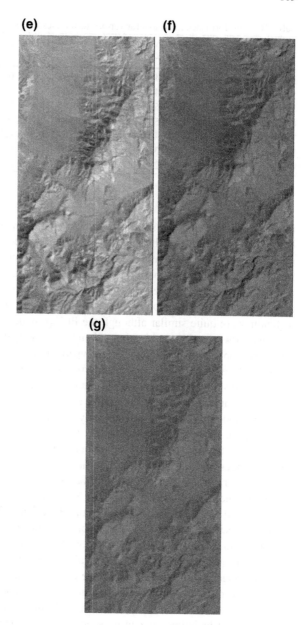

these techniques, the total amount of information transfer from the bands towards the final output is still quite high.

The coral dataset, too, is peculiar in terms of its data contents. A large percentage of the features in the coral data are visually apparent over only a very few spectral bands. Thus, features should be carefully extracted for an efficient visualization of

Table 10.5 Performance measures for various techniques for visualization of the geological data

Fusion technique	Variance σ^2	Entropy H	Avg gradient $\bar{g}$	Relative bias $\hat{b}$	Fusion factor FF	Fusion symmetry FS
Bilateral filtering technique	417.27	6.69	5.98	0.21	1.58	**0.19**
Bayesian technique	427.74	6.17	5.57	0.10	2.05	0.38
Variational technique	509.15	6.52	5.45	0.11	2.00	0.21
Optimization-based technique	**523.46**	**6.87**	**6.71**	0.19	**2.85**	0.33
Three band selection	491.07	6.70	6.54	**0.09**	1.32	0.44
Piecewise linear function	325.52	6.23	5.61	0.24	1.71	0.38
Color matching function	282.37	5.28	4.68	0.23	1.75	0.34

the scene. Figure 10.5a, c shows the results of fusion of the coral data using the bilateral filtering-based technique and the variational technique, respectively. We are able to observe certain features in the top right corner of the scene in both these images. The optimization-based fusion brings out these details, but they are not very prominent due to a relatively low contrast in color. In other results, these features although present do not appear very prominent. The results of the PLF and the CMF techniques are quite similar although the PLF provides a better discriminability of features. However, the combination of bands chosen by the band selection technique has picked up some noisy data which is visible in the form of horizontal strips. The objective assessment of the coral data from Table 10.6 indicates high amounts of contrast and sharpness in the result of the band selection technique.

This example indicates susceptibility of the no-reference measures to the noise. This technique produces good result only when a large number of features exist in the particular set of three chosen bands. This technique, however, may introduce a huge shift in the mean value of the output from that of the input data, as indicated by high values of the relative bias $\hat{b}$ for most of the datasets used in this chapter. The Bayesian, variational, and optimization-based techniques exhibit high values of the fusion symmetry (FS) that corresponds to a non-uniform participation from input hyperspectral bands towards fusion. For the Bayesian technique, the participation of a band is less if the values of the corresponding sensor selectivity factor β are small.

Finally, let us discuss the result of fusion of the urban hyperspectral data which we have used for the primary illustrations of all the techniques in the respective chapters of the monograph. In Fig. 10.6, we have provided the same results along with the results from other comparative techniques. These data contain a large number of smaller objects which are easily discriminable in the fusion result of the bilateral filtering-based technique as can be seen from Fig. 10.6a. This technique has also proved to be quite superior in the case of the Moffett Field datasets which contain similar agglomeration of smaller objects. This superiority is due to defining the fusion weights from the locally dominant features present at each pixel in each band of the data. A Bayesian solution, too, examines every individual pixel in the data, and picks up those which possess visually important information. The output of this technique provided in Fig. 10.6b is also similar to the one from the former technique. The

Fig. 10.5 Results of visualization of the coral data from the Hyperion using **a** the bilateral filtering-based technique, **b** the Bayesian technique, **c** the variational technique, **d** the optimization-based technique, **e** the three band selection technique, **f** the piecewise linear function technique, and **g** the color matching function technique

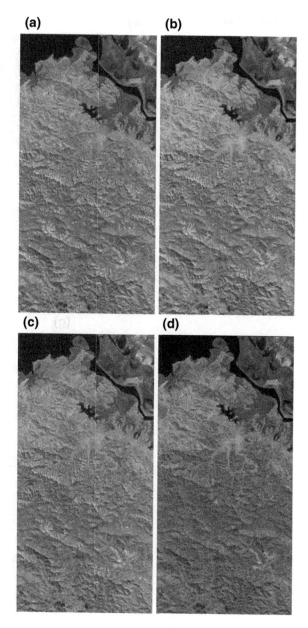

variational technique also provides a fusion result similar to that of the optimization-based technique, however, it appears less sharper as we have imposed a smoothness constraint on the fusion output. Thus, the edges and boundaries of various objects get slightly blurred. The outcome of the optimization-based technique appears visually different due to the color assignment. However, it does not reflect on the quality of

Fig. 10.5 (continued)

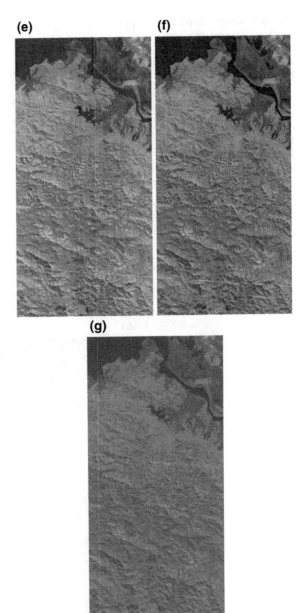

the image. Rather, this result from Fig. 10.6d brings out details such as roads, smaller objects at the central part of the image in a very clear manner.

The boundaries between these rectangular objects are also clearly visible at several places. The band selection technique has not been able to capture several such details from the data.

Table 10.6 Performance measures for various techniques for visualization of the coral data

Fusion technique	Variance σ^2	Entropy H	Avg gradient $\bar{g}$	Relative bias $\hat{b}$	Fusion factor FF	Fusion symmetry FS
Bilateral filtering technique	1195.82	6.98	6.97	0.21	1.36	0.23
Bayesian technique	**1435.03**	**7.09**	**7.07**	0.26	1.35	0.52
Variational technique	1087.80	6.91	6.73	**0.18**	1.43	0.51
Optimization-based technique	984.32	6.82	6.87	0.24	**1.65**	0.61
Three band selection	884.82	6.86	6.71	0.79	0.38	**0.14**
Piecewise linear function	871.78	6.75	5.83	0.25	1.40	0.42
Color matching function	534.30	6.00	4.70	0.24	1.34	0.19

The details of the quantitative assessment of the results of the urban data are provided in Table 10.7. We may also observe high values of the variance and the average gradient for all the presented techniques as against the corresponding values for the other techniques described here except possibly for the variational technique. The smoothness constraint in this case helps producing smooth and thus visually pleasing images, but at the cost of reduced sharpness. The contributions from the input bands are also quite high for all the discussed techniques as observed from the fusion factor *FF*. Fusion using the PLF and the CMF have captured some of the subtle features from the data, however, their results appear quite dull, and thus lack visual appeal and the clarity of details. As per Table 10.7, all techniques have performed well in terms of low values of the relative bias $\hat{b}$ with the variational technique providing the smallest value indicating the least deviation from the input data with respect to the mean intensity value. Also, low values of fusion symmetry *FS* indicate high level of uniformity in the participation from various hyperspectral bands towards the fusion output for all techniques studied in this chapter.

We now discuss the computational requirements of these fusion techniques before we conclude this chapter. We have implemented all the algorithms in MATLAB® version R2009a running on a P-IV computer running at 2.66 GHz with 2 GB RAM. Out of four techniques discussed in the monograph, the bilateral filtering-based one is a non-iterative single pass procedure. The technique initially produces a bilaterally filtered output of every band in the data. These outputs are used to compute the fusion weights while the actual fused image is produced via a normalized sum of bands. We have used the fast implementation of bilateral filter as described in [126]. The moffett$_2$, moffett$_3$, and lunar datasets provided by the AVIRIS contain 224 bands of dimensions (614×512) pixels each. For the aforementioned computer system, the bilateral filtering-based solution took nearly 120 s to generate the fused image with 3 stages of hierarchical processing which has been described in Chap. 3. The geological, coral, and urban datasets contain 242 bands of the size (512×256) pixels each, provided by the Hyperion. The proposed solution took nearly 27 s to generate the fused image with 3 stages of hierarchical processing.

The Bayesian technique can be split into two parts. The first part deals with the computation of the sensor selectivity factors β over every pixel in the data.

Fig. 10.6 Results of visualization of the urban data
from the Hyperion using **a**
the bilateral filtering-based
technique, **b** the Bayesian
technique (reproduced from
Fig. 5.2b), **c** the variational
technique, **d** the optimization-
based technique (reproduced
from Fig. 7.1b), **e** the three
band selection technique, **f**
the piecewise linear function
technique, and **g** the color
matching function technique

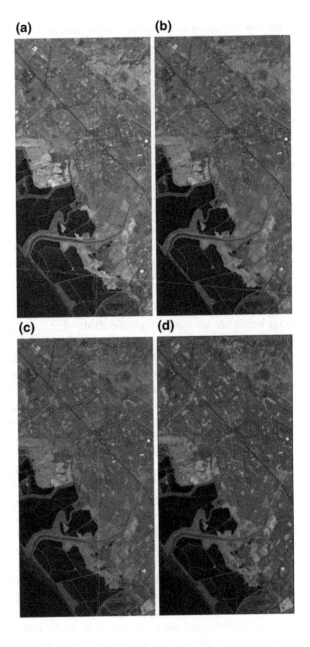

This procedure was completed in 320 s for the AVIRIS datasets, and in 98 s for the
Hyperion datasets on the same computer system described earlier. The second part
of the technique is estimation of the fused image from the so obtained β-factors.
This being an iterative procedure, the exact time for the convergence of the solution
varies with the data, and the parameters of the algorithm. The average time required

Fig. 10.6 (continued)

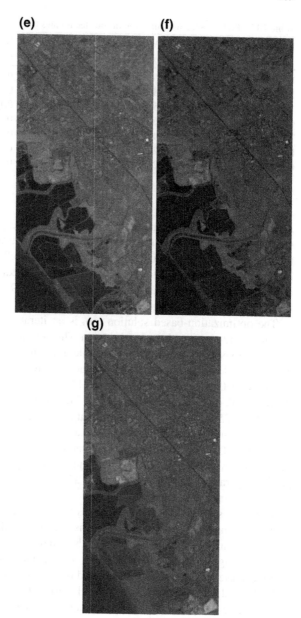

for a single iteration was 42 and 16 s for the AVIRIS and the Hyperion datasets, respectively. We found the solutions to converge in around 8–10 iterations.

Contrary to the Bayesian technique, the variational technique is a single step process. The fusion weights are not computed explicitly, but the weighting function is a part of the variational framework developed for the fusion problem. While the

Table 10.7 Performance measures for various techniques for visualization of the urban data

Fusion technique	Variance σ^2	Entropy H	Avg gradient $\bar{g}$	Relative bias $\hat{b}$	Fusion factor FF	Fusion symmetry FS
Bilateral filtering technique	619.50	6.28	6.68	0.17	1.49	**0.11**
Bayesian technique	**855.56**	6.34	7.16	0.21	**2.04**	0.16
Variational technique	558.26	6.14	6.57	**0.13**	1.58	0.26
Optimization-based technique	665.11	**6.43**	**7.46**	0.24	1.47	0.14
Three band selection	477.04	6.41	6.13	0.25	1.33	0.12
Piecewise linear function	282.54	5.80	4.55	0.31	1.34	0.22
Color matching function	260.90	5.78	4.59	0.22	1.20	0.23

fusion process involves significant computation at every iteration, it is possible to speed up the computation to some extent by pre-computing the data-dependent terms that remain unchanged throughout the fusion procedure. The average time for a single iteration for the AVIRIS data was 72 s, while the time required for the single iteration of the Hyperion images was found to be 28 s. Nearly 9–10 iterations were required for the solutions to converge.

The optimization-based solution too, is an iterative solution where the fusion weights are re-computed at every iteration. This solution, thus, turns out to be computationally expensive. Additionally, each iteration involves several terms that require a dot product and an element-wise product. On average a single iteration took 92 s for the AVIRIS datasets of dimensions $(614 \times 512 \times 224)$ pixels, while the Hyperion datasets with dimensions $(512 \times 256 \times 242)$ pixels produced the result of a single iteration in 35 s. It took nearly 6–8 iterations for all the datasets till the solutions converge. This algorithm, however, has a vector operation defined over the spectral arrays $(s(x, y))$. Each dataset contains XY spectral arrays which are processed independently to generate the pixels of the fused image. We believe that these independent operations can be parallelized to a large extent so that the computation time can be drastically reduced.

The band selection technique chooses the three bands through several filtering operations, and thus turns out to be quite fast. The PLF and the CMF techniques assign a fixed weight to each of the spectral bands, and thus the process of fusion reduces to merely a weighted linear addition where weights have already been predefined. On average, these techniques generated fused images from each of the AVIRIS datasets in 29, 12, and 11 s, respectively. For the Hyperion datasets, the average computation times required to generate the fused images for these techniques were 12, 7, and 7 s, respectively. The details of the computation time for these techniques have been summarized in Table 10.8.

Table 10.8 Average computation time requirements for various fusion techniques

Fusion technique	AVIRIS dataset (s) $(614 \times 512 \times 224)$	Hyperion dataset (s) $(512 \times 256 \times 242)$
Bilateral filtering technique	120	27
Bayesian technique	615	208
Variational technique	712	273
Optimization-based technique	740	281

10.4 Remarks

We have provided results of fusion using four different techniques discussed in the monograph, along with some other recent and commonly used techniques for visualization of hyperspectral data. We have provided the resultant fused images over several hyperspectral datasets, and also provided several quantitative performance measures for the same. From both the visual and the quantitative analysis, it can be inferred that all the presented techniques yield better quality fused images. The bilateral filtering-based technique identifies the locally dominant features in the data, and defines the fusion weights at every pixel location based on those features. The edge-preserving property of the bilateral filter helps generating the fused image without any visible artifacts. The fused image possesses a high amount of contrast and sharpness imparted by the dominant local features. This solution has low values of mean bias $\hat{b}$ as the resultant image does not deviate much from the radiometric mean of the input data. High values of the fusion factor FF, and low values of the fusion symmetry FS indicate higher and yet uniform contribution from the constituent spectral bands towards the fusion process.

The Bayesian fusion technique determines the visual quality of each pixel in the data based on two different quality factors-well-exposedness and sharpness. As the pixels with higher values of these factors contribute more towards fusion, the resultant images possess high amount of sharpness and visual clarity. However, this leads to a somewhat uneven contribution of spectral bands towards the final fused image, which results in high values of the fusion symmetry FS. The TV-norm based regularization prevents smoothening of edges and other sharp discontinuities, and yet maintains the naturalness of the fused image.

The matte-less technique combines the input hyperspectral bands without explicitly computing the fusion weights. The variational framework incorporates the weighting function as a part of the cost function. Through an iterative procedure, this technique generates a fused image with a very small deviation from intensities of the input bands. Thus, we observe very small values of the relative bias $\hat{b}$ for the variational technique. It also imposes a smoothness constraint on the fused image, which however leads to a reduction in the sharpness of the fused image, and thus, the values of the variance and average gradient are not very high in this case.

The last technique discussed in the monograph is based on optimizing the fusion weights on certain desired properties of the fused image. The well-exposedness and local contrast are considered to be the desired properties. The fused images, therefore, possess high values of variance and average gradient. However, this output-driven approach leads to a non-uniform contribution of bands and a shift in the radiometric mean of the data. Therefore, the values of the fusion symmetry FS and relative bias $\hat{b}$ are quite high for this technique.

The band selection techniques discard a large amount of information by selecting only 3 bands from 200+ spectral bands. The contents of the composite RGB image, therefore, are dependent upon the contents of the chosen bands which in turn depend upon the materials present in the scene. Hence, the performance of this technique is quite unpredictable. The PLF and the CMF techniques assign a single weight to all the pixels in a given band. These techniques, thus, do not exploit the contents of individual bands. The resultant fused images possess good contrast, sharpness, and visual quality if the bands that have been assigned higher fusion weights contain data with higher quality. Otherwise, the fused images may lack these qualities. Fusion solutions presented in the monograph exploit the entire hyperspectral data cube by computing fusion weights for each pixel in every band independently. These fusion solutions process the data on a per pixel basis which makes them better for visualization of hyperspectral data. The gain in quality of the fusion results is, however, obtained at a cost of higher computation.

Chapter 11
Conclusions and Directions for Future Research

11.1 Introduction

This monograph is a dedicated text dealing with the problem of fusion of hyperspectral data for visualization purposes. A hyperspectral image consists of a few hundred spectral bands which together encompass the reflectance response of the scene over a large bandwidth. One can think of hyperspectral image as a collection of a large number of bands that are narrow-band and contiguous in design. The visualization of this huge multidimensional data over a standard display device is an important but quite a challenging task. In this monograph, we explain various solutions to this problem in the form of visualization-oriented image fusion. Through each of the four solutions, we explore different aspects of the fusion process while the primary objective of visualization remains the same. None of the presented solutions assume any priori knowledge of the scene or the imaging system. The fused image gets generated solely from the input hyperspectral data without requiring any external information. In the next section, we provide some concluding remarks on the fusion techniques discussed in the monograph. We also discuss future directions for research in this area in Sect. 11.3.

11.2 Conclusions

The fusion techniques combine disparate information provided by the multiple spectral bands into a single composite image which is more informative than any single band in the data. The process of fusion operates over individual pixels in the data. Thus, the techniques presented in the monograph can be categorized as the pixel-based fusion schemes. The pixel-based fusion schemes typically generate the fused image through a weighted linear combination of input spectral bands where the choice of the fusion weights is the most critical part of the fusion algorithm. Hence, the weight function defines an α-matte for the purpose of combining the input bands.

S. Chaudhuri and K. Kotwal, *Hyperspectral Image Fusion*,
DOI: 10.1007/978-1-4614-7470-8_11, © Springer Science+Business Media New York 2013

The first technique is based on an edge-preserving filter known as the bilateral filter for the calculation of the α-matte. The hyperspectral bands contain a large number of objects and features. Some of the features are weak, and visually apparent over only a very small number of bands. The use of a bilateral filter enables us to extract these weak and minor features from the hyperspectral data which might get lost during the process of fusion otherwise. The fusion weights are calculated from these extracted features in order to accentuate their representation in the fused image. This technique uses the edge-preserving property of the bilateral filter in order to define the fusion weights which are functions of the locally dominant features within the bands. This technique exploits all available information at all locations in all bands, unlike the methods that assign the same weight for the entire band. This solution turns out to be very fast with the availability of faster implementations of the bilateral filter.

We have also explained a hierarchical scheme for an efficient implementation of the bilateral filtering-based solution. This scheme can accommodate any increase in the number of bands with a minimal degradation in the quality of the fused image. This implementation operates over only a fraction of data at any given instant, and therefore, does not require the entire data to be loaded into the memory. This scheme allows fusion of bands up to any given spectral bandwidth for the midband visualization which can be useful if the features from a certain set of bands need highlighting. The hierarchical scheme is generic, and therefore can be suitably applied over any pixel-based fusion techniques.

We have then developed a Bayesian framework for fusion of hyperspectral data. Here, fusion has been posed as an estimation problem where the fused image has been regarded as the *true scene* to be estimated. This fusion solution employs a model of image formation which relates the fused image with the input hyperspectral data through a first order approximation. We discussed an approach to determine the parameters of this model which indicate the closeness of the input data from the fused image to be estimated. When the fused image is meant for visualization, one expects a higher contribution towards fusion from the pixels that possess a high amount of visually important information. This fusion technique, thus, utilizes some of the characteristics of the input hyperspectral bands to determine the parameters of this model. It considers the well-exposedness and the sharpness of pixels as the quality measures for the computation of these model parameters. Like the first solution, this solution also operates on a per pixel basis in order to exploit the entire data. The fused image is then estimated using the MAP framework. The prior term in the MAP formulation has been a total variation (TV) norm-based penalty function to acknowledge the intra-pixel spatial correlation in natural images. The advantage of the TV norm-based penalty lies in its discontinuity preserving characteristic which is responsible for producing the resultant image with a high degree of sharpness. Apart from the output quality, the Bayesian fusion technique provides other advantages like flexibility in the computation of the corresponding model parameters through the choice of various image quality measures, and the availability of a large number of tools from the estimation theory.

The variational technique developed for fusion of hyperspectral images eliminates the process of generating fusion mattes. The fusion weights which act as the intermediate variables of the fusion process, define the fractional weight assigned to each pixel for the combining purposes. Thus, for each of the input images, these weights form what is known as the fusion matte. The variational fusion technique does not explicitly compute the fusion weights, and hence the fusion mattes. This solution models fusion weights as a data dependent term, and develops a weighting function as a part of the fusion algorithm. The weighting function is based on two terms—local contrast in the input data and adherence to the radiometric mean in the fused image. The local contrast which can be measured via the local variance, spatial gradient, etc., is an input-dependent (and hence a constant) term. The mean correction, however, is a quantity to be measured for the present output i.e., the fused image. We address this problem using the calculus of variations, and provide a solution using the Euler-Lagrange equation. The solution iteratively seeks to generate a resultant image that balances the radiometric property in the scene with reference to the input hyperspectral image. The variational technique penalizes the fused image for its departure from smoothness. This leads to a certain amount of blurring of the edges and boundaries in the fused image.

In the last solution we focus on certain properties of the fused image that are desirable for a better visualization. As the primary objective of fusion is visualization, the fusion process is expected to provide the best possible output image, independent of the characteristics of the input hyperspectral image. Images with a high amount of local contrast are visually appealing. However, the high contrast should not push the pixels into over- or under-saturation which reduces the information content in the image. This solution defines fusion as a multi-objective optimization problem based on these desired characteristics of the output. Like in the previous method, the final solution has been provided using the Euler-Lagrange equation. The novelty of the solution lies in defining fusion weights based on the characteristics of the fused image. This optimization-based solution has several interesting aspects as follows. In order to consider the spatial correlation among the pixels of the fused image, a commonly employed approach is to enforce a smoothness constraint over the fused image as explained in the variational fusion technique. However, this constraint often leads to over-smoothening of the output, blurring of edges and thereby degrading the visual quality of the image. The optimization-based technique acknowledges the spatial correlation within the scene by enforcing the smoothness constraint on the fusion weights rather than the fused image. The fusion weights typically have two additional constraints—the non-negativity and the normalization. We have explained how to address these constraints without transforming the optimization problem into a computationally expensive constrained optimization problem. The fused images are able to represent the scene contents with clarity and a high contrast, as desired. The optimization-based technique does not consider any of the input characteristics for the fusion process. Therefore, the solution is generic, and can be implemented for fusion of images with different modality.

While the topic of hyperspectral image fusion is being investigated, it is highly necessary to develop appropriate performance measures for the quantitative assessment of these fusion techniques. The existing measures for assessment of generalized image fusion that involves only a very few images, may not be readily suitable for assessing the fusion of hyperspectral data. We have explained modifications in several existing measures in order to extend their applicability towards an objective assessment of the fused images, and thereby the performance of the corresponding techniques. Several improvements in some of the existing measures have also been suggested which would help in a better analysis and assessment of the fusion process. The notion of *consistency* of a fusion technique has been illustrated. The consistency analysis reveals the behavior of a fusion technique as an increasing number of images are fused. We believe that this analysis would be highly useful in determining the adaptability and suitability of a generalized fusion technique for fusion of hyperspectral data. The performance measures have been categorized into three sets. The first set of measures evaluates the quality of the fused image without any reference to ground truth. The measures in the second set analyze the consistency of the fusion technique. The last set of measures investigates the contribution of each spectral band towards the final result. Thus, these measures deal with an important but often neglected aspect of the participation of each band into the fusion process. Although the main objective here has been a quantitative assessment of fusion of hyperspectral data, these performance measures can effectively be used for the assessment of generalized image fusion.

Hyperspectral image fusion deals with selectively combining nearly 200+ bands to obtain a single image representation of the scene. The fusion techniques explained in this monograph, as well as other existing techniques process each of the bands either as a whole or at a per pixel basis, for efficient fusion. The spectral bands in the hyperspectral image, however, exhibit a large amount of redundancy as the contents of the scene vary gradually over the successive hyperspectral bands. Due to a very high degree of inter-band correlation, the adjacent bands contribute a very little additional information towards the fusion result. Hence, one may select a subset of a few, but specific hyperspectral bands which collectively capture most of the information content from the input hyperspectral data. Fusion of this subset of data, thus, would be almost equivalent to fusion of the entire set of hyperspectral bands. In this monograph, we discuss two information theoretic schemes for band selection based on redundancy elimination. The first scheme selects a subset of the hyperspectral bands that are mutually less correlated from each other. The process of band selection, thus involves computation of the conditional entropy of the given hyperspectral band with respect to each of the previously selected bands. As the hyperspectral bands are usually ordered according to the wavelengths, we can assume a specific model to define the conditional entropy as a function of the spectral distance between the bands. Under this assumption, the process of band selection reduces to the computation of conditional entropy of the given band with respect to the last selected band only. Along with visual illustrations for the quality of fused image over a reduced subset, we have also provided the bounds on the savings in computation. This scheme is independent of the fusion process, and depends only on the properties of the input hyperspectral data. We have also discussed another band selection scheme based on the fusion technique to be applied. In this scheme, the subset of selected hyperspectral bands is fused to generate an intermediate fused image. A hyperspectral band is included in this subset only when it is significantly less correlated with this intermediate fused image obtained prior to inclusion of this band. The band selection process now involves computation of the intermediate fused images whenever a new input band is selected. This output-based band selection turns out to be computationally expensive. Unlike the first method of band selection, it selects only those hyperspectral bands that contain significant additional information as compared to the existing fusion output. For both band selection schemes, there is a very minimal degradation in the quality of the resultant fused images even when one selects less than 25 % of the original data.

11.3 Future Directions

The focus of work in this monograph has been on fusion of hyperspectral images and the performance evaluation of the fusion algorithm. We have explained in detail various fusion techniques and each method has been shown to have its own merits and demerits. Despite dealing with different useful topics in this monograph, the research in hyperspectral image visualization is far from being complete. There are several interesting aspects which can be considered for further research.

- The edge preserving solution uses a bilateral filter for calculation of fusion weights. The bilateral filter has two parameters to control the amount of filtering in the range and spatial domains, respectively. The current solution assumes the values of these parameters in a heuristic way. One would like to calculate these values from the data in an adaptive manner. The values of the bilateral kernel parameters should be based on the statistical and spatial properties of the input data.
- Several improvements to the bilateral filter have been suggested, for example a trilateral filter [41]. It would be worth exploring whether such improvements would help in extracting the minor features in the data in a better manner. Further, anisotropic diffusion has been shown to be closely related to the mathematical definition of bilateral filtering [11, 12]. Hence, one may explore the usefulness of anisotropic diffusion instead of a bilateral filter.
- The Bayesian solution involves a model for image formation. The computation of the parameters of this model is based on multiple quality measures of a given pixel. While our experiments have considered a couple of such measures, it will be interesting to experiment with various other measures of image quality. The set of computationally simple measures that can capture the visual information most efficiently should be identified.
- If a high value of performance measure is the key goal, it will also be interesting to evaluate the performances of various desired quality measures by embedding them directly into the cost function of the optimization-based solution. Since our fusion solutions are intended for human visualization, designing of the performance criteria functions that incorporate some the human visual system (HVS)-based quality measures would be quite useful and relevant.
- The Bayesian and the optimization-based techniques explicitly consider the spatial correlation, i.e., the intra-band correlation. The consecutive bands of the

hyperspectral data also exhibit a very degree of correlation among them, i.e., the spectral or the inter-band correlation. We have not explicitly considered this factor which may lead to some further improvement in the results.

- The variational technique employs the local variance measure for computation of contrast at a pixel level. One may want to replace it by other measures such as spatial gradient which may be more effective, and faster in computation.

- We have discussed a framework for the consistency analysis of a fusion technique, and demonstrated the same through experimentation. However, the task of mathematically proving that a given fusion technique is indeed consistent still remains. Based on the consistency analysis, we would like to develop a fusion technique that will be consistent with respect to a given performance measure. With a wide choice of performance measures available, these experiments might provide a variety of solutions along with some interesting insights towards the concept of image fusion. It should also be noted that since the result of fusion depends on the input data, proving mathematically that a fusion technique is consistent for a given performance measure for all possible datasets (i.e., data independence) will be very difficult. If an appropriate statistical model of the hyperspectral data cube is assumed, it may be possible to prove the consistency of a fusion scheme mathematically.

- In hyperspectral imaging, the spectral signatures of various types of materials on the earth surface are often available. Would such a prior knowledge help in the fusion process? Our understanding is that it may be possible to use such information for a better quality image fusion. However, this would require a segmented map of the region on the earth surface being imaged, when one may have to combine the task of scene classification and image fusion into a single problem definition.

- Since hyperspectral data cube has a very large number of image bands with a high amount of correlation along the spectral axis, we have discussed two different ways by which a large number of constituent bands can be eliminated while fusing the data. However, both these approaches are greedy in nature. The first band is trivially selected. It is quite possible that, for a given fraction of the total number of bands selected, a different combination of image bands may provide a better visualization. It may be worthwhile to explore how one can obtain the solution that provides the best visualization.

- In this monograph several different methods of image fusion have been discussed, and their performances have been compared. While closing, it is quite natural to ask which specific method is being prescribed to solve the general problem of hyperspectral image fusion? It is, indeed, quite difficult to answer this question. None of the methods yields the best scores for all performance measures defined in this monograph. However, it does appear from the results given in the previous chapter that the optimization-based technique tends to yield better scores for a majority of these performance measures. However, it involves more computations which may limit its applicability for fast visualization of the hyperspectral data cube. Fortunately, a large amount of such computations can be done in parallel. It would be interesting to have this technique, and possibly other techniques too, implemented on a GPU to speed up the computation.

References

1. Aiazzi, B., Alparone, L., Baronti, S., Garzelli, A.: Context-driven fusion of high spatial and spectral resolution images based on oversampled multiresolution analysis. IEEE Trans. Geosci. Remote Sens. **40**(10), 2300–2312 (2002)
2. Aiazzi, B., Baronti, S., Selva, M.: Improving component substitution pansharpening through multivariate regression of MS + Pan data. IEEE Trans. Geosci. Remote Sens. **45**(10), 3230–3239 (2007)
3. Allen, E., Triantaphillidou, S.: The Manual of Photography. Elsevier, Oxford (2010)
4. Alparone, L., Facheris, L., Baronti, S., Garzelli, A., Nencini, F.: Fusion of multispectral and SAR images by intensity modulation. In: Proceedings of International Conference on Information Fusion, pp. 637–643, Stockholm, Sweden (2004)
5. Amolins, K., Zhang, Y., Dare, P.: Wavelet based image fusion techniques: an introduction, review and comparison. ISPRS J. Photogram. Remote Sens. **62**(4), 249–263 (2007)
6. Ardouin, J., Levesque, J., Rea, T.: A demonstration of hyperspectral image exploitation for military applications. In: Proceedings of International Conference on Information Fusion, pp. 1–8, Québec, Canada (2007)
7. Asmare, M., Asirvadam, V., Iznita, L.: Multi-sensor image enhancement and fusion for vision clarity using contourlet transform. In: Proceedings of International Conference on Information Management and Engineering, pp. 352–356, Kuala Lumpur, Malaysia (2009)
8. Aubert, G., Deriche, R., Kornprobst, P.: Computing optical flow via variational techniques. SIAM J. Appl. Math. **60**(1), 156–182 (1999)
9. Avcibas, I., Bülent, S., Khalid, S.: Statistical evaluation of image quality measures. J. Electron. Imaging **11**(2), 206–223 (2002)
10. The AVIRIS website. http://aviris.jpl.nasa.gov (2012). Accessed Jan 2012
11. Barash, D.: Fundamental relationship between bilateral filtering, adaptive smoothing, and the nonlinear diffusion equation. IEEE Trans. Pattern Anal. Mach. Intell. **24**(6), 844–847 (2002)
12. Barash, D., Comaniciu, D.: A common framework for nonlinear diffusion, adaptive smoothing, bilateral filtering and mean shift. Image Vis. Comput. **22**(1), 73–81 (2004)
13. Bennett, E.P., McMillan, L.: Video enhancement using per-pixel virtual exposures. ACM Trans. Graph. **24**(3), 845–852 (2005)
14. Black, M., Sapiro, G.: Edges as outliers: anisotropic smoothing using local image statistics. In: Proceedings of Scale-Space Theories in Computer Vision, vol. 1682, pp. 259–270. Springer, Berlin (1999)
15. Blum, R.S.: Robust image fusion using a statistical signal processing approach. Inf. Fusion **6**(2), 119–128 (2005)
16. Bogoni, L.: Extending dynamic range of monochrome and color images through fusion. In: Proceedings of International Conference on Pattern Recognition, vol. 3, pp. 7–12, Barcelona, Spain (2000)

17. Borman, S.: Topics in multiframe superresolution restoration. Ph.D. thesis, University of Notre Dame (2004)

18. Boykov, Y., Huttenlocher, D.: A new Bayesian framework for object recognition. In: Proceedings of Conference on Computer Vision and Pattern Recognition, vol. 2, pp. 637–663, Ft. Collins, USA (1999)

19. Buades, A., Coll, B., Morel, J.M.: A non-local algorithm for image denoising. In: Proceedings of IEEE Computer Society Conference on Computer Vision and Pattern Recognition, vol. 2, pp. 60–65, San Diego, USA (2005)

20. Burt, P., Kolczynski, R.: Enhanced image capture through fusion. In: Proceedings of International Conference on Computer Vision, pp. 173–182, Berlin, Germany (1993)

21. Burt, P.J.: The Pyramid as a Structure for Efficient Computation. Rensselaer Polytechnic Institute, Troy (1983)

22. Cai, S., Du, Q., Moorhead, R.: Hyperspectral imagery visualization using double layers. IEEE Trans. Geosci. Remote Sens. 45(10), 3028–3036 (2007)

23. Candès, E.J., Donoho, D.: Curvelets: a surprisingly effective nonadaptive representation for objects with edges. In: Proceedings of International Conference on Curves and Surfaces, vol. 2, pp. 1–7, San Malo, France (1999)

24. Cao, W., Li, B., Zhang, Y.: A remote sensing image fusion method based on PCA transform and wavelet packet transform. In: Proceedings of the International Conference on Neural Networks and Signal Processing, vol. 2, pp. 976–981, Nanjing, China (2003)

25. Carmona, R., Zhong, S.: Adaptive smoothing respecting feature directions. IEEE Trans. Image Process. 7(3), 353–358 (1998)

26. Chai, Y., He, Y., Ying, C.: CT and MRI image fusion based on contourlet using a novel rule. In: Proceedings of International Conference on Bioinformatics and Biomedical Engineering, pp. 2064–2067, Shanghai, China (2008)

27. Chambolle, A., Lions, P.L.: Image recovery via total variation minimization and related problems. Numer. Math. 76(2), 167–188 (1997)

28. Chan, T.F., Golub, G.H., Mulet, P.: A nonlinear primal-dual method for total variation-based image restoration. SIAM J. Sci. Comput. 20(6), 1964–1977 (1999)

29. Chang, C.I. (ed.): Hyperspectral Data Exploitation: Theory and Applications, 1st edn. Wiley-Interscience, Hoboken (2007)

30. Chaudhuri, S.: Super-resolution Imaging. Kluwer Academic, Boston (2001)

31. Chaudhuri, S., Joshi, M.V.: Motion-free Super-resolution. Springer, New York (2005)

32. Chaudhury, K., Sage, D., Unser, M.: Fast O(1) bilateral filtering using trigonometric range kernels. IEEE Trans. Image Process. 20(12), 3376–3382 (2011)

33. Chavez, P., Sises, S., Anderson, J.: Comparison of three different methods to merge multiresolution and multispectral data: Landsat TM and SPOT panchromatic. Photogram. Eng. Remote Sens. 57(3), 295–303 (1991)

34. Chen, K.: Adaptive smoothing via contextual and local discontinuities. IEEE Trans. Pattern Anal. Mach Intell. 27(10), 1552–1567 (2005)

35. Chen, T., Guo, R., Peng, S.: Image fusion using weighted multiscale fundamental form. In: Proceedings of International Conference on Image Processing, vol. 5, pp. 3319–3322, Singapore (2004)

36. Chen, T., Zhang, J., Zhang, Y.: Remote sensing image fusion based on ridgelet transform. In: Proceedings of International Geoscience and Remote Sensing Symposium, vol. 2, pp. 1150–1153, Seoul, Korea (2005)

37. Chen, Y., Xue, Z., Blum, R.: Theoretical analysis of an information-based quality measure for image fusion. Inf. Fusion 9(2), 161–175 (2008)

38. Cho, M., Skidmore, A., Corsi, F., van Wieren, S.E., Sobhan, I.: Estimation of green grass/herb biomass from airborne hyperspectral imagery using spectral indices and partial least squares regression. Int. J. Appl. Earth Obs. Geoinf. 9(4), 414–424 (2007)

39. Choi, M., Kim, R., Kim, M.: The curvelet transform for image fusion. In: Proceedings of Congress of the International Society for Photogrammetry and Remote Sensing, vol. B8, pp. 59–64, Istanbul, Turkey (2004)

40. Choi, M., Kim, R.Y., Nam, M.R., Kim, H.O.: Fusion of multispectral and panchromatic satellite images using the curvelet transform. IEEE Geosci. Remote Sens. Lett. **2**(2), 136–140 (2005)
41. Choudhury, P., Tumblin, J.: The trilateral filter for high contrast images and meshes. In: Proceedings of ACM SIGGRAPH 2005 Courses, Los Angeles, USA (2005)
42. Clark, M.L., Roberts, D.A., Clark, D.B.: Hyperspectral discrimination of tropical rain forest tree species at leaf to crown scales. Remote Sens. Environ. **96**(3–4), 375–398 (2005)
43. Comaniciu, D., Meer, P.: Mean shift: a robust approach toward feature space analysis. IEEE Trans. Pattern Anal. Mach. Intell. **24**(5), 603–619 (2002)
44. Cui, M., Razdan, A., Hu, J., Wonka, P.: Interactive hyperspectral image visualization using convex optimization. IEEE Trans. Geosci. Remote Sens. **47**(6), 1673–1684 (2009)
45. Dasarathy, B.V.: Fusion strategies for enhancing decision reliability in multisensor environments. Opt. Eng. **35**(3), 603–616 (1996)
46. Daubechies, I.: Ten lectures on wavelets. Society for Industrial and Applied Mathematics, Philadelphia (1992)
47. De, I., Chanda, B.: A simple and efficient algorithm for multifocus image fusion using morphological wavelets. Signal Process. **86**(5), 924–936 (2006)
48. D'Elia, C., Poggi, G., Scarpa, G.: A tree-structured markov random field model for bayesian image segmentation. IEEE Trans. Image Process. **12**(10), 1259–1273 (2003)
49. Demir, B., Çelebi, A., Ertürk, S.: A low-complexity approach for the color display of hyperspectral remote-sensing images using one-bit-transform-based band selection. IEEE Trans. Geosci. Remote Sens. **47**(1), 97–105 (2009)
50. Do, M., Vetterli, M.: The contourlet transform: an efficient directional multiresolution image representation. IEEE Trans. Image Process. **14**(12), 2091–2106 (2005)
51. Dobson, D.C., Santosa, F.: Recovery of blocky images from noisy and blurred data. SIAM J. Appl. Math. **56**(4), 1181–1198 (1996)
52. Donoho, D.L., Duncan, M.R.: Digital curvelet transform: strategy, implementation, and experiments. In: Szu, H., Vetterli, M., Campbell, W., Buss, J. (eds.) Proceedings of Aerosense, pp. 12–30. VII, Wavelet Applications vol (2000)
53. Dozier, J., Painter, T.: Multispectral and hyperspectral remote sensing of alpine snow properties. Annu. Rev. Earth Planet. Sci. **32**, 465–494 (2004)
54. Durand, F., Dorsey, J.: Fast bilateral filtering for the display of high-dynamic-range images. ACM Trans. Graph. **21**(3), 257–266 (2002)
55. Elad, M.: On the origin of the bilateral filter and ways to improve it. IEEE Trans. Image Process. **11**(10), 1141–1151 (2002)
56. Erives, H., Targhetta, N.: Implementation of a 3-D hyperspectral instrument for skin imaging applications. IEEE Trans. Instrum. Meas. **58**(3), 631–638 (2009)
57. Farsiu, S., Robinson, M., Elad, M., Milanfar, P.: Fast and robust multiframe super resolution. IEEE Trans. Image Process. **13**(10), 1327–1344 (2004)
58. Fattal, R., Agrawala, M., Rusinkiewicz, S.: Multiscale shape and detail enhancement from multi-light image collections. ACM Trans. Graph. **26**(3), 51:1–51:9 (2007)
59. Fleishman, S., Drori, I., Cohen-Or, D.: Bilateral mesh denoising. ACM Trans. Graph. **22**(3), 950–953 (2003)
60. Forster, B., Van De Ville, D., Berent, J., Sage, D., Unser, M.: Complex wavelets for extended depth-of-field: a new method for the fusion of multichannel microscopy images. Microsc. Res. Tech. **65**(1–2), 33–42 (2004)
61. Garzelli, A., Nencini, F., Alparone, L., Aiazzi, B., Baronti, S.: Pan-sharpening of multispectral images: a critical review and comparison. In: Proceedings of International Geoscience and Remote Sensing Symposium, vol. 1, pp. 81–84, Alaska, USA (2004)
62. Garzelli, A., Nencini, F., Capobianco, L.: Optimal MMSE pan-sharpening of very high resolution multispectral images. IEEE Trans. Geosci. Remote Sens. **46**(1), 228–236 (2008)
63. Geman, S., Geman, D.: Stochastic relaxation, gibbs distributions, and the bayesian restoration of images. IEEE Trans. Pattern Anal. Mach. Intell. **6**(6), 721–741 (1984)

64. Golub, G., Heath, M., Wahba, G.: Generalized cross-validation as a method for choosing a good ridge parameter. Technometrics **21**, 215–223 (1979)

65. Gonzalez, R., Woods, R.: Digital Image Processing, 2nd edn. Pearson Education, New Jersey (2002)

66. Gonzalez-Audicana, M., Saleta, J., Catalan, R., Garcia, R.: Fusion of multispectral and panchromatic images using improved IHS and PCA mergers based on wavelet decomposition. IEEE Trans. Geosci. Remote Sens. **42**(6), 1291–1299 (2004)

67. Goshtasby, A.A.: Fusion of multi-exposure images. Image Vis. Comput. **23**(6), 611–618 (2005)

68. Goshtasby, A.A., Nikolov, S.: Image fusion: advances in the state of the art. Inf. Fusion **8**(2), 114–118 (2007)

69. Gowen, A., O'Donnell, C., Cullen, P., Downey, G., Frias, J.: Hyperspectral imaging: an emerging process analytical tool for food quality and safety control. Trends Food Sci. Technol. **18**(12), 590–598 (2007)

70. Greig, D., Porteous, B., Seheult, A.: Exact maximum a posteriori estimation for binary images. J. Roy. Stat. Soc. Ser. B (Methodol.) **51**(2), 271–279 (1989)

71. Guo, B., Gunn, S., Damper, B., Nelson, J.: Hyperspectral image fusion using spectrally weighted kernels. In: Proceedings of International Conference on Information Fusion, vol. 1, pp. 1–7, PA, USA (2005)

72. Hardie, R., Eismann, M., Wilson, G.: MAP estimation for hyperspectral image resolution enhancement using an auxiliary sensor. IEEE Trans. Image Process. **13**, 1174–1184 (2004)

73. Herman, G., De Pierro, A., Gai, N.: On methods for maximum a posteriori image reconstruction with a normal prior. J. Vis. Commun. Image Represent. **3**(4), 316–324 (1992)

74. Hewson, R., Cudahy, T., Caccetta, M., Rodger, A., Jones, M., Ong, C.: Advances in hyperspectral processing for province- and continental-wide mineral mapping. In: Proceedings of IEEE International Geoscience and Remote Sensing Symposium, vol. 4, pp. 701–704, Cape Town, South, Africa (2009)

75. Hong, S., Sudduth, K., Kitchen, N., Drummond, S., Palm, H., Wiebold, W.: Estimating within-field variations in soil properties from airborne hyperspectral images. In: Proceedings of Pecora 15/Land Satellite Information IV Conference, pp. 1–13, Denver, USA (2002)

76. Horn, B., Brooks, M.J.: Shape from Shading. MIT Press, Cambridge (1989)

77. The hyperion website. http://eo1.usgs.gov (2012). Accessed Jan 2012

78. Jacobson, N., Gupta, M.: Design goals and solutions for display of hyperspectral images. IEEE Trans. Geosci. Remote Sens. **43**(11), 2684–2692 (2005)

79. Jacobson, N., Gupta, M., Cole, J.: Linear fusion of image sets for display. IEEE Trans. Geosci. Remote Sens. **45**(10), 3277–3288 (2007)

80. Jain, A.K.: Fundamentals of Digital Image Processing. Prentice Hall, Upper Saddle River (1989)

81. Jiang, W., Baker, M., Wu, Q., Bajaj, C., Chiu, W.: Applications of a bilateral denoising filter in biological electron microscopy. J. Struct. Biol. **144**(1–2), 114–122 (2003)

82. Jiji, C.V., Joshi, M.V., Chaudhuri, S.: Single-frame image super-resolution using learned wavelet coefficients. Int. J. Imaging Syst. Technol. **14**(3), 105–112 (2004)

83. Joshi, M., Bruzzone, L., Chaudhuri, S.: A model-based approach to multiresolution fusion in remotely sensed images. IEEE Trans. Geosci. Remote Sens. **44**(9), 2549–2562 (2006)

84. Kailath, T.: The divergence and Bhattacharyya distance measures in signal selection. IEEE Trans. Commun. Technol. **15**(1), 52–60 (1967)

85. Kass, M., Witkin, A., Terzopoulos, D.: Snakes: active contour models. Int. J. Comput. Vis. **1**(4), 321–331 (1988)

86. Kirankumar, Y., Devi, S.S.: Transform-based medical image fusion. Int. J. Biomed. Eng. Technol. **1**(1), 101–110 (2007)

87. Kotwal, K., Chaudhuri, S.: A fast approach for fusion of hyperspectral images through redundancy elimination. In: Proceedings of Indian Conference on Computer Vision, Graphics and Image Processing, ICVGIP '10, pp. 506–511, Chennai, India (2010)

88. Kotwal, K., Chaudhuri, S.: Visualization of hyperspectral images using bilateral filtering. IEEE Trans. Geosci. Remote Sens. **48**(5), 2308–2316 (2010)
89. Kotwal, K., Chaudhuri, S.: An optimization-based approach to fusion of multi-exposure, low dynamic range images. In: Proceedings of International Conference on Information Fusion (FUSION), vol. 1, pp. 1–7, Chicago, USA (2011)
90. Kotwal, K., Chaudhuri, S.: An optimization-based approach to fusion of hyperspectral images. IEEE J. Sel. Top. Appl. Earth Obs. Remote Sens. **5**(2), 501–509 (2012)
91. Kotwal, K., Chaudhuri, S.: A Bayesian approach to visualization-oriented hyperspectral image fusion. Inf. Fusion (2013). (accepted)
92. Kotwal, K., Chaudhuri, S.: A novel approach to quantitative evaluation of hyperspectral image fusion techniques. Inf. Fusion **14**(1), 5–18 (2013)
93. Kruse, F., Boardman, J., Huntington, J.: Comparison of airborne hyperspectral data and EO-1 Hyperion for mineral mapping. IEEE Trans. Geosci. Remote Sens. **41**(6), 1388–1400 (2003)
94. Kumar, M.: Total variation regularization-based adaptive pixel level image fusion. In: Proceedings of IEEE Workshop on Signal Processing Systems, pp. 25–30, San Francisco, USA (2010)
95. Kumar, M., Dass, S.: A total variation-based algorithm for pixel-level image fusion. IEEE Trans. Image Process. **18**(9), 2137–2143 (2009)
96. Laliberte, F., Gagnon, L., Sheng, Y.: Registration and fusion of retinal images- an evaluation study. IEEE Trans. Med. Imaging **22**(5), 661–673 (2003)
97. Laliberte, F., Gagnon, L., Yunlong, S.: Registration and fusion of retinal images: a comparative study. In: Proceedings of International Conference on Pattern Recognition, vol. 1, pp. 715–718, Québec, Canada (2002)
98. Lambrecht, R., Lambrecht, R., Woodhouse, C.: Way Beyond Monochrome 2e: Advanced Techniques for Traditional Black & White Photography Including Digital Negatives and Hybrid Printing. Elsevier, Oxford (2010)
99. Lee, J.S.: Digital image enhancement and noise filtering by use of local statistics. IEEE Trans. Pattern Anal. Mach. Intell. **2**(2), 165–168 (1980)
100. Lee, S.J., Rangarajan, A., Gindi, G.: Bayesian image reconstruction in SPECT using higher order mechanical models as priors. IEEE Trans. Med. Imaging **14**(4), 669–680 (1995)
101. Leonardis, A., Bischof, H., Pinz, A. (eds.): European Conference on Computer Vision, part I, vol. 3951. Springer, Heidelberg (2006)
102. Levin, A., Lischinski, D., Weiss, Y.: Colorization using optimization. ACM Trans. Graph. **23**(3), 689–694 (2004)
103. Lewis, J.J., O'Callaghan, R.J., Nikolov, S.G., Bull, D.R., Canagarajah, N.: Pixel- and region-based image fusion with complex wavelets. Inf. Fusion **8**(2), 119–130 (2007)
104. Li, H., Manjunath, B.S., Mitra, S.: Multisensor image fusion using the wavelet transform. Graph. Models Image Process. **57**(3), 235–245 (1995)
105. Liang, P., Wang, Y.: Local scale controlled anisotropic diffusion with local noise estimate for image smoothing and edge detection. In: Proceedings of International Conference on Computer Vision, pp. 193–200, Bombay, India (1998)
106. Lin, J.: Divergence measures based on the Shannon entropy. IEEE Trans. Inf. Theory **37**(1), 145–151 (1991)
107. Liu, Z., Tsukada, K., Hanasaki, K., Ho, Y., Dai, Y.: Image fusion by using steerable pyramid. Pattern Recogn. Lett. **22**(9), 929–939 (2001)
108. Lu, P.Y., Huang, T.H., Wu, M.S., Cheng, Y.T., Chuang, Y.Y.: High dynamic range image reconstruction from hand-held cameras. In: Proceedings of IEEE Conference on Computer Vision and Pattern Recognition, pp. 509–516, Miami, USA (2009)
109. Mahmood, Z., Scheunders, P.: Enhanced visualization of hyperspectral images. In: IEEE International Geoscience and Remote Sensing Symposium (IGARSS), pp. 991–994, Hawaii, USA (2010)
110. Malin, D.F.: Unsharp masking. AAS Photo Bull. **16**, 10–13 (1977)
111. Mallat, S.: A Wavelet Tour of Signal Processing. Elsevier Science, Oxford (1999)

112. Melas, D., Wilson, S.: Double Markov random fields and Bayesian image segmentation. IEEE Trans. Signal Process. **50**(2), 357–365 (2002)
113. Mertens, T., Kautz, J., Van Reeth, F.: Exposure fusion. In: Proceedings of Pacific Conference on Computer Graphics and Applications, pp. 382–390, Hawaii, USA (2007)
114. Miao, Q., Wang, B.: A novel image fusion method using contourlet transform. In: Proceedings of International Conference on Communications, Circuits and Systems, vol. 1, pp. 548–552, Guilin, China (2006)
115. Mignotte, M.: A multiresolution Markovian fusion model for the color visualization of hyperspectral images. IEEE Trans. Geosci. Remote Sens. **48**(12), 4236–4247 (2010)
116. Mitianoudis, N., Stathaki, T.: Pixel-based and region-based image fusion schemes using ICA bases. Inf. Fusion **8**(2), 131–142 (2007)
117. Mitri, G., Gitas, I.: Mapping postfire vegetation recovery using EO-1 Hyperion imagery. IEEE Trans. Geosci. Remote Sens. **48**(3), 1613–1618 (2010)
118. Moeller, M., Wittman, T., Bertozzi, A.L.: A variational approach to hyperspectral image fusion. In: Society of Photo-Optical Instrumentation Engineers (SPIE) Conference Series, Society of Photo-Optical Instrumentation Engineers (SPIE) Conference Series, vol. 7334 (2009)
119. Moghaddam, B., Jebara, T., Pentland, A.: Bayesian face recognition. Pattern Recogn. **33**(11), 1771–1782 (2000)
120. Murray, D.W., Kashko, A., Buxton, H.: A parallel approach to the picture restoration algorithm of Geman and Geman on an SIMD machine. Image Vis. Comput. **4**(3), 133–142 (1986)
121. Nolin, A., Dozier, J.: A hyperspectral method for remotely sensing the grain size of snow. Remote Sens. Environ. **74**(2), 207–216 (2000)
122. Osher, S., Sole, A., Vese, L.: Image decomposition and restoration using total variation minimization and the H sup 1. Multiscale Model. Simul. **1**(3), 349–370 (2003)
123. van Ouwerkerk, J.D.: Image super-resolution survey. Image. Vis. Comput. **24**(10), 1039–1052 (2006)
124. Pajares, G., Manuel, J.: A wavelet-based image fusion tutorial. Pattern Recogn. **37**(9), 1855–1872 (2004)
125. Paolo, N.: Variational approach to optical flow estimation managing discontinuities. Image Vis. Comput. **11**(7), 419–439 (1993)
126. Paris, S., Durand, F.: A fast approximation of the bilateral filter using a signal processing approach. In: Proceedings of the European Conference on Computer Vision, pp. 568–580, Graz, Austria (2006)
127. Park, S., Park, M., Kang, M.G.: Super-resolution image reconstruction: a technical overview. IEEE Signal Process. Mag. **20**(3), 21–36 (2003)
128. Peli, E.: Contrast in complex images. J. Opt. Soc. Am. **7**(10), 2032–2040 (1990)
129. Peng, G., Ruiliang, P., Biging, G., Larrieu, M.: Estimation of forest leaf area index using vegetation indices derived from hyperion hyperspectral data. IEEE Trans. Geosci. Remote Sens. **41**(6), 1355–1362 (2003)
130. Perona, P., Malik, J.: Scale-space and edge detection using anisotropic diffusion. IEEE Trans. Pattern Anal. Mach. Intell. **12**(7), 629–639 (1990)
131. Petrović, V., Xydeas, C.: Multiresolution image fusion using cross band feature selection. Proc. SPIE **3719**, 319–326 (1999)
132. Petrović, V., Xydeas, C.: Gradient-based multiresolution image fusion. IEEE Trans. Image Process. **13**(2), 228–237 (2004)
133. Petschnigg, G., Szeliski, R., Agrawala, M., Cohen, M., Hoppe, H., Toyama, K.: Digital photography with flash and no-flash image pairs. In: ACM SIGGRAPH, pp. 664–672, Los Angeles, California (2004)
134. Pham, T.Q., van Vliet, L.J.: Separable bilateral filtering for fast video preprocessing. In: Proceedings of IEEE International Conference on Multimedia and Expo, pp. 1–4, Amsterdam, The Netherlands (2005)
135. Piella, G.: A region-based multiresolution image fusion algorithm. In: Proceedings of International Conference on Information Fusion, vol. 2, pp. 1557–1564, Washington DC, USA (2002)

136. Piella, G.: A general framework for multiresolution image fusion: from pixels to regions. Inf. Fusion **4**(4), 259–280 (2003)
137. Piella, G.: Image fusion for enhanced visualization: a variational approach. Int. J. Comput. Vis. **83**(1), 1–11 (2009)
138. Piella, G., Heijmans, H.: A new quality metric for image fusion. In: Proceedings of International Conference on Image Processing, vol. 3, pp. 173–176, Barcelona, Spain (2003)
139. Poggio, T., Koch, C.: Ill-posed problems in early vision: from computational theory to analogue networks. Proc. Roy. Soc. London. Ser. B, Biol. Sci. **226**(1244), 303–323 (1985)
140. Pohl, C., Van Genderen, J.L.: Review article multisensor image fusion in remote sensing: concepts, methods and applications. Int. J. Remote Sens. **19**(5), 823–854 (1998)
141. Polesel, A., Ramponi, G., Mathews, V.: Image enhancement via adaptive unsharp masking. IEEE Trans. Image Process. **9**(3), 505–510 (2000)
142. Porikli, F.: Constant time O(1) bilateral filtering. In: Proceedings of IEEE Conference on Computer Vision and Pattern Recognition, pp. 1–8, Anchorage, USA (2008)
143. Qi, J., Leahy, R., Cherry, S.R., Chatziioannou, A., Farquhar, T.H.: High-resolution 3D Bayesian image reconstruction using the microPET small-animal scanner. Phys. Med. Biol. **43**(4), 1001 (1998)
144. Qu, G., Zhang, D., Pingfan, Y.: Information measure for performance of image fusion. Electron. Lett. **38**(7), 313–315 (2002)
145. Raman, S., Chaudhuri, S.: A matte-less, variational approach to automatic scene compositing. In: Proceedings of International Conference on Computer Vision, pp. 1–6, Rio De Janeiro, Brazil (2007)
146. Raman, S., Chaudhuri, S.: Bilateral filter based compositing for variable exposure photography. In: Proceedings of EuroGraphics Conference, pp. 1–4, Munich, Germany (2009)
147. Ramesh, C., Ranjith, T.: Fusion performance measures and a lifting wavelet transform based algorithm for image fusion. In: Proceedings of International Conference on Information Fusion, vol. 1, pp. 317–320, Washington DC, USA (2002)
148. Ramponi, G., Strobel, N., Mitra, S., Yu, T.H.: Nonlinear unsharp masking methods for image contrast enhancement. J. Electron. Imaging **5**(3), 353–366 (1996)
149. Ranchin, T., Aiazzi, B., Alparone, L., Baronti, S., Wald, L.: Image fusion: the ARSIS concept and some successful implementation schemes. ISPRS J. Photogram. Remote Sens. **58**(1–2), 4–18 (2003)
150. Ready, P., Wintz, P.: Information extraction, SNR improvement, and data compression in multispectral imagery. IEEE Trans. Commun. **21**(10), 1123–1131 (1973)
151. Redondo, R., Šroubek, F., Fischer, S., Cristòbal, G.: Multifocus image fusion using the log-gabor transform and a multisize windows technique. Inf. Fusion **10**(2), 163–171 (2009)
152. Ren, H., Chang, C.I.: Automatic spectral target recognition in hyperspectral imagery. IEEE Trans. Aerosp. Electron. Syst. **39**(4), 1232–1249 (2003)
153. Rodger, A., Lynch, M.: Determining atmospheric column water vapour in the 0.4–2.5 μm spectral region. In: Proceedings of the 10th JPL Airborne Earth Science, Workshop, pp. 321–330 (2001)
154. Roy, V., Goita, K., Granberg H. and Royer, A.: On the use of reflective hyperspectral remote sensing for the detection of compacted snow. In: Proceedings of International Geoscience and Remote Sensing Symposium (IGARSS), pp. 3263–3266, Denver, USA (2006)
155. Rudin, L.I., Osher, S., Fatemi, E.: Nonlinear total variation based noise removal algorithms. Physica D **60**(1–4), 259–268 (1992)
156. Saint-Marc, P., Chen, J.S., Medioni, G.: Adaptive smoothing: a general tool for early vision. IEEE Trans. Pattern Anal. Mach. Intell. **13**(6), 514–529 (1991)
157. Salem, F., Kafatos, M.: Hyperspectral image analysis for oil spill mitigation. Proceedings of Asian Conference on Remote Sensing, Singapore, In (2001)
158. Scheunders, P.: A multivalued image wavelet representation based on multiscale fundamental forms. IEEE Trans. Image Process. **11**(5), 568–575 (2002)
159. Scheunders, P., De Backer, S.: Fusion and merging of multispectral images with use of multiscale fundamental forms. J. Opt. Soc. Am. A **18**(10), 2468–2477 (2001)

160. Schowengerdt, R.: Remote Sensing: Models and Methods for Image Processing, 3rd edn. Academic, New York (2007)
161. Schultz, R.A., Nielsen, T., Zavaleta, J., Ruch, R., Wyatt, R., Garner, H.: Hyperspectral imaging: a novel approach for microscopic analysis. Cytometry **43**(4), 239–247 (2001)
162. Shah, V.P., Younan, N.H., King, R.L.: An efficient pan-sharpening method via a combined adaptive PCA approach and contourlets. IEEE Trans. Geosci. Remote Sens. **46**(5), 1323–1335 (2008)
163. Sharma, R.: Probabilistic model-based multisensor image fusion. Ph.D. thesis, Oregon Institute of Science and Technology, Portland, Oregon (1999)
164. Sharma, R., Leen, T.K., Pavel, M.: Bayesian sensor image fusion using local linear generative models. Opt. Eng. **40**(7), 1364–1376 (2001)
165. Shaw, G., Manolakis, D.: Signal processing for hyperspectral image exploitation. IEEE Signal Process. Mag. **19**(1), 12–16 (2002)
166. Sifakis, E., Garcia, C., Tziritas, G.: Bayesian level sets for image segmentation. J. Vis. Commun. Image Represent. **13**(1–2), 44–64 (2002)
167. Simoncelli, E., Freeman, W.: The steerable pyramid: a flexible architecture for multi-scale derivative computation. In: Proceedings of International Conference on Image Processing, vol. 3, pp. 444–447, Washington DC, USA (1995)
168. Smith, R.: Introduction to hyperspectral imaging. Technical report, Microimages Inc (2012)
169. Smith, S.M., Brady, J.M.: SUSAN: a new approach to low level image processing. Int. J. Comput. Vis. **23**(1), 45–78 (1997)
170. Strachan, I.B., Pattey, E., Boisvert, J.B.: Impact of nitrogen and environmental conditions on corn as detected by hyperspectral reflectance. Remote Sens. Environ. **80**(2), 213–224 (2002)
171. Tatzer, P., Wolf, M., Panner, T.: Industrial application for inline material sorting using hyperspectral imaging in the NIR range. Real-Time Imaging **11**(2), 99–107 (2005)
172. Toet, A.: Hierarchical image fusion. Mach. Vis. Appl. **3**(1), 1–11 (1990)
173. Toet, A., Franken, E.M.: Perceptual evaluation of different image fusion schemes. Displays **24**(1), 25–37 (2003)
174. Toet, A., Toet, E.: Multiscale contrast enhancement with applications to image fusion. Opt. Eng. **31**, 1026–1031 (1992)
175. Tomasi, C., Manduchi, R.: Bilateral filtering for gray and color images. In: Proceedings of International Conference on Computer Vision, pp. 839–846, Bombay, India (1998)
176. Tsagaris, V., Anastassopoulos, V., Lampropoulos, G.: Fusion of hyperspectral data using segmented PCT for color representation and classification. IEEE Trans. Geosci. Remote Sens. **43**(10), 2365–2375 (2005)
177. Tyo, J., Konsolakis, A., Diersen, D., Olsen, R.: Principal-components-based display strategy for spectral imagery. IEEE Trans. Geosci. Remote Sens. **41**(3), 708–718 (2003)
178. Vaidyanathan, P.P.: Multirate Systems and Filter Banks. Pearson Education, London (1993)
179. Vo-Dinh, T.: A hyperspectral imaging system for in vivo optical diagnostics. IEEE Eng. Med. Biol. Mag. **23**(5), 40–49 (2004)
180. Wald, L.: Some terms of reference in data fusion. IEEE Trans. Geosci. Remote Sens. **37**(3), 1190–1193 (1999)
181. Wang, H., Peng, J., Wu, W.: Fusion algorithm for multisensor images based on discrete multiwavelet transform. IEE Proc. Vis. Image Signal Process. **149**(5), 283–289 (2002)
182. Wang, Q., Shen, Y., Zhang, Y., Zhang, J.: A quantitative method for evaluating the performances of hyperspectral image fusion. IEEE Trans. Instrum. Meas. **52**(4), 1041–1047 (2003)
183. Wang, Z., Bovik, A.: A universal image quality index. IEEE Signal Process. Lett. **9**(3), 81–84 (2002)
184. Wang, Z., Bovik, A., Sheikh, H., Simoncelli, E.: Image quality assessment: from error visibility to structural similarity. IEEE Trans. Image Process. **13**(4), 600–612 (2004)
185. Wang, Z., Bovik, A.C., Lu, L.: Why is image quality assessment so difficult? In: Proceedings of International Conference on Acoustics, Speech, and Signal Processing, vol. 4, pp. 3313–3316, Florida, USA (2002)

186. Wang, Z., Sheikh, H., Bovik, A.: No-reference perceptual quality assessment of JPEG compressed images. In: Proceedings of International Conference on Image Processing, vol. 1, pp. 477–480, New York, USA (2002)

187. Weiss, B.: Fast median and bilateral filtering. ACM Trans. Graph. **25**(3), 519–526 (2006)

188. Wen, C., Chen, J.: Multi-resolution image fusion technique and its application to forensic science. Forensic Sci. Int. **140**(2–3), 217–232 (2004)

189. Wilson, T., Rogers, S., Kabrisky, M.: Perceptual-based image fusion for hyperspectral data. IEEE Trans. Geosci. Remote Sens. **35**(4), 1007–1017 (1997)

190. Wong, A.: Adaptive bilateral filtering of image signals using local phase characteristics. Signal Process. **88**(6), 1615–1619 (2008)

191. Wu, W., Yao, J., Kang, T.: Study of remote sensing image fusion and its application in image classification. In: Proceedings of Commission VII ISPRS Congress, vol. XXXVII, p. 1141. International Society for Photogrammetry and Remote Sensing, Beijing, China (2008)

192. Xu, M., Chen, H., Varshney, P.K.: A novel approach for image fusion based on Markov random fields. In: Proceedings of Annual Conference on Information Sciences and Systems, pp. 344–349, Princeton, USA (2008)

193. Xu, M., Chen, H., Varshney, P.K.: An image fusion approach based on Markov random fields. IEEE Trans. Geosci. Remote Sens. **49**(12), 5116–5127 (2011)

194. Xydeas, C., Petrović, V.: Objective image fusion performance measure. Electron. Lett. **36**(4), 308–309 (2000)

195. Yang, B., Jing, Z.: Medical image fusion with a shift-invariant morphological wavelet. In: Proceedings of IEEE Conference on Cybernetics and Intelligent Systems, pp. 175–178, Chengdu, China (2008)

196. Yang, J., Blum, R.: A statistical signal processing approach to image fusion for concealed weapon detection. In: Proceedings of International Conference on Image Processing, vol. 1, pp. 513–516, New York, USA (2002)

197. Yang, L., Guo, B.L., Ni, W.: Multimodality medical image fusion based on multiscale geometric analysis of contourlet transform. Neurocomputing **72**(1–3), 203–211 (2008)

198. Yang, Q., Tan, K.H., Ahuja, N.: Real-time O(1) bilateral filtering. In: Proceedings of IEEE Conference on Computer Vision and Pattern Recognition, pp. 557–564, Miami, USA (2009)

199. Yang, Q., Yang, R., Davis, J., Nister, D.: Spatial-depth super resolution for range images. In: Proceedings of IEEE Conference on Computer Vision and Pattern Recognition, pp. 1–8. Minneapolis, USA (2007)

200. Yuan, L., Sun, J., Quan, L., Shum, H.Y.: Image deblurring with blurred/noisy image pairs. ACM Trans. Graph. 26(3) (2007)

201. Zhang, Q., Guo, B.L.: Multifocus image fusion using the nonsubsampled contourlet transform. Signal Process. **89**(7), 1334–1346 (2009)

202. Zhang, Y., Hong, G.: An IHS and wavelet integrated approach to improve pan-sharpening visual quality of natural colour IKONOS and QuickBird images. Inf. Fusion **6**(3), 225–234 (2005)

203. Zhang, Z.: Investigations of image fusion. Ph.D. thesis, Lehigh University, PA, USA (1999)

204. Zheng, Q., Chellappa, R.: Estimation of illuminant direction, albedo, and shape from shading. In: Proceedings of Conference on Computer Vision and Pattern Recognition, pp. 540–545, Hawaii, USA (1991)

205. Zheng, Y., Essock, E., Hansen, B., Haun, A.: A new metric based on extended spatial frequency and its application to DWT based fusion algorithms. Inf. Fusion **8**(2), 177–192 (2007)

206. Zhu, Y., Varshney, P., Chen, H.: Evaluation of ICA based fusion of hyperspectral images for color display. In: Proceedings of International Conference on Information Fusion, pp. 1–7, Québec, Canada (2007)

207. Zhukov, B., Oertel, D., Lanzl, F., Reinhackel, G.: Unmixing-based multisensor multiresolution image fusion. IEEE Trans. Geosci. Remote Sens. **37**(3), 1212–1226 (1999)

208. Zurita-Milla, R., Clevers, J.: Schaepman, M.:Unmixing-based landsat TM and MERIS FR data fusion. IEEE Geosci. Remote Sens. Lett. **5**(3), 453–457 (2008)

Index

S. Chaudhuri and K. Kotwal, *Hyperspectral Image Fusion*,
DOI: 10.1007/978-1-4614-7470-8, © Springer Science+Business Media New York 2013

Printed in the United States
By Bookmasters